AN OIL GEOLOGIST ABROAD
Exploration With Family
in
Bolivia, Spain, and Nigeria
1956–1966

AN OIL GEOLOGIST ABROAD

Exploration With Family
in
Bolivia, Spain, and Nigeria
1956–1966

Eric Ericson and Libby Ericson

SANTA FE

© 2011 by Eric Ericson and Libby Ericson.
All Rights Reserved.

No part of this book may be reproduced in any form or by any electronic or mechanical means including information storage and retrieval systems without permission in writing from the publisher, except by a reviewer who may quote brief passages in a review.

Sunstone books may be purchased for educational, business, or sales promotional use. For information please write: Special Markets Department, Sunstone Press, P.O. Box 2321, Santa Fe, New Mexico 87504-2321.

Book and Cover design › Vicki Ahl
Body typeface › Bell MT
Printed on acid free paper

Library of Congress Cataloging-in-Publication Data

Ericson, Eric, 1928-
  An oil geologist abroad : exploration with family in Bolivia, Spain, and Nigeria, 1956-1966 / by Eric Ericson and Libby Ericson.
      p. cm.
  ISBN 978-0-86534-824-0 (softcover : alk. paper)
 1. Ericson, Eric, 1928---Travel. 2. Ericson, Libby, 1928—Travel. 3. Petroleum geologists--United States--Biographpy. 4. Petroleum--Prospecting--Bolivia. 5. Petroleum--Prospecting--Spain. 6. Petroleum--Prospecting--Nigeria. I. Ericson, Libby, 1928- II. Title.
  TN869.E75 2011
  622'.182820922--dc23
                    2011018438

WWW.SUNSTONEPRESS.COM
SUNSTONE PRESS / POST OFFICE BOX 2321 / SANTA FE, NM 87504-2321 /USA
(505) 988-4418 / ORDERS ONLY (800) 243-5644 / FAX (505) 988-1025

For John, Paul, and Mark

CONTENTS

Preface 9

Part I
BOLIVIA
1956–1960

1 Eric: Shreveport to Bolivia. 13
2 *Libby: Shreveport to New York* *16*
3 Eric: Road to Charagua. 20
4 *Libby: New York to Lima, Peru* *24*
5 Eric: Mandeapecua Camp. 27
6 *Libby: Lima to Cochabamba with Frank* *31*
7 Eric: New Office, Typhoid, and Hepatitis. 34
8 *Libby: Home?.* *38*
9 Eric: Asuncion de Guaryos 44
10 *Libby: Augustina and Hepatitis.* *48*
11 Eric: Missionaries and Ayoreo Indians. 53
12 *Libby: Move and Mark* *58*
13 Eric: Airplanes and a Volcano 64
14 *Libby: Vacation and Fishing* *68*
15 Eric: Santa Cruz Office 74
16 *Libby: Stateside, Astronauts and Rio De Janeiro* . *76*
17 Eric: Drilling, and on to New York. 80
18 *Libby: Santa Cruz, Farewell Bolivia* *84*

Part II
SPAIN
1960–1963

19 Eric: Colorado Rockies, and Hello Spain! 91

20 Libby: Madrid and Vitoria 96
21 Eric: Spanish and German Work-mates 102
22 Libby: The Neighborhood and Bullfights 107
23 Eric: Castillo I and Cueva de Altamira 113
24 Libby: Our Cabin and Spanish Life 118
25 Eric: Travels in Spain 125
26 Libby: Travels with the Boys, and David 131
27 Eric: Vitoria I Blows in, on to Nigeria 137
28 Libby: Isabel, Joan, and Marianistas 143

Part III
NIGERIA
1963–1966

29 Eric: Steamy Lagos 151
30 Libby: Federal Palace Hotel and the Fever 156
31 Eric: Ikoyi Club and Okan I! 160
32 Libby: Mini Minor, Schools, and JFK 165
33 Eric: Okan Field, Sarumi, and Snakes 172
34 Libby: Shopping, Femmi, and School 178
35 Eric: Oil Production, Ju Ju, and Sarumi's Party . . . 185
36 Libby: Paul's Emergency and Teef Man 191
37 Eric: Robertkiri Oil Field and Cameroon 196
38 Libby: David's Village, Mark's Party, and Loompi . . 203
39 Eric: Revolution and Transfer 208
40 Libby: Biafran War and Going Home 213

Epilogues 215

Preface

When our daughter-in-law and son, Mark, presented us with a beautiful granddaughter a few years ago, soon to be followed by another, I began to think about recording our legacy. Our lives had been extraordinary at times and needed to be clearly documented. I put aside my painting and began to dig through letters, old writings, diaries, and compile my stories on the computer. However, it soon became obvious that half of our life, and the reason for our travels was missing.

I convinced Eric to join me and I became his secretary on the computer for three summers in our mountain home in Colorado. He poured through old business records, maps and letters, and constantly amazed me with his memory of names and places as well as his professional and historical memory.

Pat Carr, Richard Anderson, Will Gates, Jerry Hotchkiss, and Sharon Stine, all published writers, encouraged our efforts, as well as our relatives and friends in Santa Fe and Colorado. We are most grateful to all for their support and confidence in our abilities at this late stage of life. However, we could not have finished this effort without the technical abilities and devotion of our son, Paul. We decided to publish the years from 1956 to 1966, concluding that they would be the most interesting to readers.

—Libby Ericson

Part I
BOLIVIA
1956–1960

1. Eric: Shreveport to Bolivia

I accepted a job after graduation from the University of Colorado in 1951 as a beginning geologist with Continental Oil Company in Tyler, Texas. I would have preferred geological field work in the Rockies but most of the class wanted the same and there weren't enough jobs to go around. I was lucky to have been offered a job with the US Geological Survey, however, I eventually wanted to work overseas. There was active drilling and subsurface exploration work down south and we newlyweds, Libby and I, were eager to leave Boulder and begin a career.

I was soon doing well site work in the vicinity of North America's largest oil accumulation at that time, the East Texas Field, and in three years I was working in a larger office in Shreveport, Louisiana. I was a Yankee from New York and Libby had lived in the Rocky Mountains since she was twelve and the variety of insects and often oppressive heat and humidity plus the racism that existed then was not the environment where we wanted to raise our family. Libby's uncle and brother-in-law, both successful geologists, participated in foreign oil exploration and we saw no reason why we couldn't do the same. My father had also worked with an international oil company and I met and envied geologists with extensive overseas experience. At the end of World War II, at 17, I joined the Marine Corps and hoped to travel but ended up at Floyd Bennett Field Naval Air Station in New York! So much for joining the Marine Corps to see the world.

I asked Continental for a transfer but they wanted me to stay in the States and work in the Gulf Coast. Libby and I decided I would go to the 1956 AAPG (American Association of Petroleum Geologists) Convention in Chicago and check out the foreign job market. On my return to Shreveport I was called by Gus Pyre, the Vice President of Latin American exploration for Gulf Oil Company who knew my father and had heard of my interests while I was in Chicago. He offered me a job with an operation just beginning in Bolivia. To a large extent, it would be starting over, but the opportunity sounded irresistible to both of us. After finding out that several couples with children were going, some of whom we had heard about, and with the encouragement of several acquaintances in this company, Libby and I decided we would take our two young sons overseas with Gulf.

Things moved fast. I was to go as soon as possible and I hated to leave Libby with so much packing yet to finish, but fortunately, our house sold immediately. The Gulf International Office had not operated in Bolivia before but had other foreign operations. All our travel arrangements, our passports, permissions to reside in Bolivia, necessary health procedures for life in a semitropical country, would all be coordinated by the New York office. All future American and other foreign employees had to have yellow fever, tetanus, typhoid and paratyphoid shots and go through rigorous physicals. The Bolivian government also required legal clearance by police departments for all foreign workers.

My first overseas plane flight began in New York with Panagra (Pan American Grace) with stops in Miami, Panama, Ecuador and finally, Lima, Peru. There we changed to a smaller prop plane scheduled for La Paz, Bolivia, that connected with Bolivian Airlines on a flight to Cochabamba where we would live. This was winter in August with crisp mountain air and bright sun.

I was met by the Company's accountant and chauffeur, and driven to a staff house on the edge of town that served as living quarters and a temporary office, called Casa Roca. One geological field party was in the field already and another was being organized for the southern part of one of our large concession areas. I was to join this group that was led by an experienced French Swiss field geologist, Jacques Jaccard, who had just arrived from Venezuela and Gulf operations there. I met with another newly arrived American geologist, Warren Souder, and a few American Gulf office personnel to help prepare us for a six weeks trip to the eastern Andean foothills and the Chaco region. The party would consist of Jacques, Warren, myself, and three local drivers, who would double as mechanics, named Luis, Carmello and Victor. Later we would pick up a cook in Santa Cruz.

We were to travel in three vehicles. A 6x6 truck and a Dodge power wagon to carry our equipment and provisions as well as a jeep for us geologists. During this time we were all beginning to know each other and we found the Bolivians were anticipating our trip as much as we were. These Bolivians came from the mountain areas of the country and were unfamiliar with the lowland regions. Luis was the eldest and I was the youngest of this international crew who were now setting off into the forest, sometimes jungle, or *selva*, as it was called in Bolivia.

While in New York, I had tried to find books about Bolivia but found only two that were descriptive of the country. One written by a French traveler in eastern Bolivia entitled *Green Hell* and the other written about an Englishman who had made surveys for the Bolivian Government in northeastern Bolivia, entitled *Exploration Fawcett.* Fawcett disappeared on his last journey but his descriptions of the jungle gave rise to a legendary figure. Although I was looking forward to the adventure, I was left a bit anxious after reading these two books and their stories of huge snakes and wild Indians.

2. Libby: Shreveport to New York

I could find Bolivia on the maps, but Chochabamba? What a funny name. The International Gulf office in New York City hadn't much information and there was no help in Shreveport, Louisiana. At that time, long distance phone calls made in the States were expensive and cablegrams were the only way to communicate with South America. I dug out my old Colorado University Spanish grammar books and our first purchases were two small Spanish dictionaries. We poured over maps in any spare time and we discovered that Bolivia was the size of Texas, New Mexico and about half of Colorado combined, with more than half of the country mountainous, peaks reaching over 21,000 feet, and a population of four million people.

Although I had lived in Tampa Florida until I was twelve, I had lived through World War II in Boulder, Colorado during my early teens, and we were constantly hearing news and personal stories about Europe, Africa and the Pacific Islands. Boulder and the CU campus were never the same. We were taken over by ROTC's, V-12's and students in the Japanese Language School. I had two brother-in-laws who were pilots during World War II, one boy friend who served in the Pacific and another who fought in the Korean War. The world welcomed our country's expanding interests after the war and the idea of living in foreign countries didn't seem quite so unusual to my generation as perhaps it would be today.

Eric was expected to leave immediately after the sale of our house. We would not be together again for probably three months, the time the company decided was enough to determine whether Eric was compatible with a job in a foreign country. However we had decided it must work and that a change in our future was necessary. It was very difficult for me to see him leave but our three and one-half year old, John, and one and one-half year old, Paul, our two beautiful little towheads, couldn't understand how long three months would be, and just kissed their daddy good by. Although we had made most decisions together about what to take, I was left to supervise packing for the long trip and our two year contract in Bolivia. The company would not pay for any return trips during that time. We would be allowed to take some furniture along with all the possessions we considered important and that should have been a warning to me that they didn't expect much to be available in Bolivia!

Our possessions would be in storage until the company was certain that Eric was right for the job. Then they would be shipped by train and boat and eventually overland from Bolivia's free port in Chile and finally by rail. Unfortunately, even though the dishes arrived unbroken, everything was damp and arrived nine months later. With help from a dear aunt who had recently moved to Shreveport and from neighbors and friends, I somehow managed to pull it all together.

The boys and I were put on a train for Denver by friends and after long visits with relatives and friends in Boulder, we proceeded on to Kansas City for a few weeks with my mother. I had two older sisters who had lived in foreign countries. One followed her geologist husband to India and the other joined the Red Cross as a psychologist and wound up in Japan after WWII. So my dear mother was stoic about her daughters traveling into foreign countries. She would yet see another daughter whose engineer husband took her to the Philippines and Australia. I got a strong okay message from all of these older adventurous sisters. My mother sent me with mostly smiles and much love to New York City on my first airplane flight. I was twenty-seven.

Our John was never able to sit still and always led me a merry chase but Paul was thankfully a more contented soul. I had visited my in-laws several times in the past five years, including on our honeymoon, which was an exciting adventure for me and included Jimmy Durante, the Rockettes and New Year's Eve with Guy Lombardo. Eric's parents had also visited us in Colorado and we'd made several vacation trips to New York with the boys but the prospect of spending a couple of months with them and without Eric was daunting. I needn't have feared. Eric's mother was a saint, just like mine. And they kept their worries to themselves about this adventure of ours.

The boys must have been a big handful for my in-laws but they took care of them while I shopped in the Big Apple or came with me if I had to take the boys into the city. They also accompanied me to far off places like the Port Authority in lower Manhattan by the piers for our yellow fever shots. Eric's sister, Joan, a bit younger, but a true New Yorker, took me all over the town for evening entertainment. I learned how to travel the subways and finally, how to tip the cab drivers enough. We even managed to find a Spanish tutor who would come for lessons at night and not just try to teach me more Spanish vocabulary but help prepare me for a life with children in a country that spoke another language.

I corresponded with two geologists' wives with small children who were also going to Bolivia and had previously lived in Spanish speaking countries. I bought the food blender they recommended, the food grinder, the pressure cooker and the absolutely necessary penicillin ointment. I even tried to buy a two-year supply of shoes and clothes for the boys. But when Dorothy Truitt suggested I buy a two-year supply of toilet paper, the logistics blew my mind and I decided to forget it. If I had known that we would be tearing Time magazine pages in half, crunching them up and skewering them on a hook next to the toilet in my fancy bathroom next to a bathtub full of brown water for almost a year, I might have rethought my decision.

Eric wrote at least weekly, and I wrote more often on our aerogrammes. I tried to keep him abreast of the activities of the boys, who missed their father so much and always thought he would be at the next stop. He tried to tell me what would be important to bring to make our life more comfortable and he sounded very happy and excited with Bolivia and his experiences there. My questions were never-ending but unfortunately he was in the jungle most of the time and really had no idea what life in the big town of Cochabamba was all about.

As all of our purchases began to pile up in my in-laws entry, my father-in-law and I watched in amazement, and could hardly believe that the company would send us and all that baggage on an airplane to a place called Cochabamba, Bolivia. The New York Gulf International office had been very welcoming and helpful to me and when I talked to the office personnel from VP's to secretaries I got nothing but encouragement. Several men in the office confided that they wished their wives would have taken on an adventure such as this. Even today, in a cynical world, I think they were sincere.

Finally, finally, we were given the green light to go. The boys and I, with all our baggage, were taken to the Idlewild airport but only we adults were tearful. Johnny and Pauly could hardly wait to see their daddy.

3. Eric: Road to Charagua

Bolivia had an agrarian reform revolution in 1952 which had distributed the large estates amongst the *campesinos* (local farmers) and nationalized the tin mines but privatized the oil industry in much of the country. The only oil field still producing, which was discovered by Standard Oil of New Jersey (Esso) in the 1930s, was at Camiri in the southern Andean foothills. Shortly after their discovery the partially developed field was taken over by the Bolivian Government during the Chaco War, which involved the western boundary between Bolivia and Paraguay.

Their new petroleum legislation opened vast areas in the eastern part of Bolivia, ranging from the Amazon Basin in the north to the Chaco plains and forest to the south. Very few attempts had been made to discover oil in that part of Bolivia and Gulf had obtained the first and largest exploration permits under the law in unmapped and unexplored territory.

Luis, Caramello and Victor would make several trips with me in the next two years. They became my instructors on local customs, food, language, and political and racial differences that existed between the Spanish and indigenous peoples. The people in the eastern part of the country, or *Oriente*, were quite different from the mountain people in their customs, dress, diet and politics. Also the climate was dramatically different with pronounced wet and dry seasons. On this trip we could only travel comfortably during the dry season and some areas were only accessible during that time.

Luis and Caramello were primarily Spanish from the intermediate mountain area of the Quechua speaking people, while Victor was from the Aymara speaking people in the Altiplano, the 12,000 foot mountain valley between 20,000 foot peaks. I soon learned the cultural differences between those peoples and began to understand how the climate and geography could affect the lowlands in the *Oriente*. In our small group it became obvious to us foreigners, although embarrassing, that since we were educated, we were included in what they considered to be the 'upper-class'. I was always called *seenyore* or *Docktoor Ehreeksown* and treated with respect. Luis, the senior driver was mild mannered but authoritative, and deferred to by the other drivers for his experience and ability. These energetic men looked out for us and helped us adjust in every way to our new environment.

We arrived in Santa Cruz, the largest city in eastern Bolivia, a city with no paved streets and a sloth hanging from a palm tree in the central plaza. The three-hundred mile trip from Cochabamba had taken two days over a newly constructed road, mostly paved by Point Four (US AID). We found a cook named Antonio who was pretty good at preparing rice, beans and a passable bread. Any fresh meat, fruit and vegetables would consist of what we could shoot or buy along the way.

While in Santa Cruz, we had been advised to stay at the Panagra guest house which was part of the weather-worn airport buildings. What sticks in my mind was a sign in the bathroom by the toilets with instructions: "*No pise sobre los asientos*", meaning 'don't stand on the toilet seats', because local public facilities were generally a hole in the floor. The manager of the guest house and restaurant, an Italian from Brazil, Mario, and his charming wife, both spoke English and attempted to keep the place up to European standards for the clientele traveling to and from eastern Bolivia and Brazil.

We were soon on our way and had to negotiate dirt roads which were barely passable and rivers with no bridges where we were attacked by dense swarms of biting black gnats whenever we stopped. There was a railroad under construction between the Argentine border and Santa Cruz, which paralleled the unpaved and sandy road that we traveled, and the first river at which we arrived was deep enough to require a ferry. Waiting our turn, it took several hours to cross and we were fresh meat for the *marajui*, the black gnats. With the enormous inflation rate of the Boliviano it took a bundle of bills to get our three vehicles to the other side. We stopped in a village for our first night and found only hammocks to sleep in, a first experience for most of us. Fortunately, we had our mosquito nets. The locals were happy to have our company and money.

Crossing the Rio Grande.

Warren and I were helpless without our multi-lingual party chief, Jacques, who could speak English with us and Spanish with the drivers and locals. We drove the next day to the small village of Charagua where we rented a house for ourselves and crew for a week. As you can imagine, this small village of perhaps twenty families, a hundred miles from the closest town, was fascinated with these *estranjeros* (strangers), their instruments and vehicles. We awakened our first morning to find a gang of kids watching our every movement intently and they laughed at Warren's and my attempts to speak Spanish.

Our house was little more than an adobe shell with a roof that would undoubtedly leak in the rainy season. However, we were able to set up cots and Antonio managed with a kerosene stove to feed us. There was no electricity or running water so we read by candlelight or kerosene lantern and our outdoor privy had a partial adobe wall with a hole in the ground.

We surveyed the area from the town site east to the Rio Parapeti that would establish a baseline to tie in with the aerial photography that was being flown at this time. Before leaving, we geologists were invited by the local *Alcalde* (Mayor) for dinner at his home. We brought beer, Don Francisco had some Chilean wine, and we dined on his patio on very tough but tasty chicken served by a barefooted girl. For dessert an open can of sweet peaches was brought to the table. Besides the bats flying around, the most fascinating thing to Warren and me was a little flag in the middle of the table signifying that our host was a member of Rotary International.

4. Libby: New York to Lima, Peru

The flight was memorable from beginning to end, from NYC to Cochabamba, Bolivia. I found myself between NYC and Washington DC with several rather impressive Wall Street young dads on their way to DC, and some onto Miami, all who missed their own families and very eager to share parenting problems with this young blonde mother from out West. The boys were handed around and entertained but when they began to fuss for food, the stewardesses informed us that dinner would not be served until after we left DC. That meant that we had to entertain them with crackers until 8:30 PM and the stewardesses said they couldn't even bring milk! My father-in-law had called Panagra before leaving for Idlewild and was assured that there would be no problem with food or milk for the boys.

Few women and children traveled by air in 1956. It seemed to be the world of business men and these guys were quite upset along with me when 'their little boys' couldn't be fed. Some men left the plane at DC but the remaining were as shocked as I was to find out that since John was half price and Paul was flying with no charge, Paul could have no food and John only a half portion. And since John ate almost as much as me at the time, yours truly had mostly crackers for dinner.

I dressed my tired little boys in their PJ's and bedded them down for the long overnight flight to Lima, Peru. We were scheduled to arrive sometime the next morning and all that had happened in the last three months began to catch up with me. I had a glass of wine with my men friends and was discussing this adventure of Eric's and mine, which they thought was insane of course, when Miami was announced. The stewardess told us that we must all disembark in Miami. When I explained to them that I had not been informed previously that we must leave the plane, that the boys were asleep, and I didn't want to wake them, they ignored me.

The men were indignant. Several went up to discuss the situation with the pilots and the others said they would report this incident when they disembarked in Miami. At this point I was close to tears, but the stewardesses were adamant that I must leave the plane while we were in Miami. We landed and almost immediately both pilots showed up and assured me that I most certainly did not have to disembark! The new stewardesses that came aboard were charming and along with the new passengers were Latin Americans, bilingual and very supportive of the boys and me. I fell asleep and I remember nothing until we arrived in Lima.

Friends of one of my sisters and her husband whom I had known while we were in university now lived in Lima, and had been advised of our arrival. Dorothy Wirtz met our plane and carried us off to her beautiful home. It was surrounded by high white plastered walls, covered with climbing red flowering vines, and her three children, a bit older than mine, completely took over the boys. Their father was a mining engineer and at a mine high in the mountains but the five children and we two mothers had a lovely afternoon, with many instructions about life in a Spanish speaking country from all the family. Dot even insisted I tuck a Peruvian pottery antiquity in Paul's diaper bag which she was sure no customs agent would venture into. Upon arrival in Cochabamba, Eric couldn't believe I had done such a thing, considering we had never even dared smuggle out a piece of petrified wood while he was a Ranger at Dinosaur National Monument the summer after we married.

Dinner was mostly prepared by the maid and at last to bed, for an early rise and our morning flight to Bolivia. However, in the night I was awakened from a dead sleep when her nine-year old boy cried out. I was closest to his room and when I peeped out my door into a pitch black hall, I heard running. It seemed someone had climbed over the wall, entered the house and made ghost noises in his bedroom.

Dot called an American neighbor and soon big, black booted Peruvian policemen were stomping around, shouting Spanish and brandishing guns. I stood watching and listening, trying to understand even one word for several hours. I'm sure I made their night, watching this newly arrived American woman huddled in a blanket and open-mouthed half the time. It turned out that the intruder was just the local looney that everyone knew about and was caught down the street in another house before the police left Dot's house. Eventually there were smiles between all and I finally understood the friendly *hasta luegos*.

5. Eric: Mandeapecua Camp

Our ultimate destination on this trip was the old, abandoned Mandeapequa drilling camp of YPFB (*Yacimientos Petroliferos Fiscales de Bolivia*), the state oil company, where an unsuccessful exploratory test had been drilled. It was on the most eastern surface fold on the Chaco plain and our job was to learn the stratigraphy and rework the surface geology. We arrived at Mandeapecua after a difficult crossing of the Rio Parapeti, where we had to winch the trucks through the sandy ford. We found a few crumbling structures with partial roofs where we decided to set up our camp for possibly several months. Our nearest neighbors, as far as we knew, would be in Boyuibe which was along the railroad line to Argentina. They would provide us with some supplies of canned foods, chickens and eggs but no fresh vegetables or fruit.

    We organized a water filtration system with a barrel into which we could pump water from an old well for drinking and showers, and then we dug a latrine. Several rainy days gave us an opportunity to become more acquainted and reflect on our project as well as our new environment. During our trip Jacques had explained to us some of the health problems we might encounter. One of these was the deadly Chagas disease which was carried in the feces of the South American version of our 'kissing bug'. With some relish, he told us that the disease was not easily detected and if carried several years in your system, could result in death.

    We had set up mosquito netting but shortly after our arrival, I awakened to find several of these 'kissing bugs' in the furthest corner inside my mosquito netting. Although I could find no marks on my body, they looked frighteningly well fed, so I captured and put them in a box to be examined. All of us were concerned because old buildings could be infested with them. We sent Caramello to the town of Camiri, about eighty miles over dirt roads to the YPFB headquarters, where they had an infirmary in their Operations Center. We had been advised of these facilities in case we needed help of any kind. He returned a day later with the news that I must wait several weeks for the results and after four anxious weeks, they reported that the 'kissing bugs' showed no sign of the Chagas parasite. It didn't completely reassure me and in the back of my mind I waited for years for the maximum time to pass for the disease to become fatal. Little did I know then, but found out much later, that a world renowned Brazilian specialist in Chagas disease had been visiting Camiri to do more study on the Bolivian variety of the parasites and that he had personally cleared mine.

The good old jeep.

We had heard from the locals that on the east side of the Parapeti after it turned north and began to flow into the sand channels that the wild Indians had been seen. This was confirmed by the *Alcalde* of Charagua. These Indians, Ayoreos, or 'Barbaros' as they were called, had been known to kill missionaries and locals and we would be working at the edge of the Indians' range. Our job was to survey the outcrops in the vicinity of Mandeapequa and map any structures we could identify.

We wanted to begin with the outcrops in the canyon of the Rio Parapeti to the northwest and we found a good rock section in the river. I had never imagined a scrub forest that could be so dry, dense, and with such thorny brush. Now we also met with a bee slightly bigger than the gnats, very appropriately named the 'sweat bee', which did not sting but swarmed over any sweaty clothes or body parts and made life miserable.

We had several firearms in our field party, including a .22 and 30/30 carbine and a 12 gauge shotgun. These were for hunting any game we could eat as well as protect ourselves. In the 1930s when Standard Oil was working in the region, there were several incidents where bandits had robbed the camps of their payrolls. Opportunities to shoot game were rare and we saw very little wildlife in the area. We did shoot several strange birds as well as doves which were a welcome change from the tough chicken and canned Argentinian corned beef. Before leaving we were visited by a European rancher with whom we exchanged hunting stories. He was not happy that we had killed one particular bird that acted as a 'guard' bird to warn people of approaching strangers. We felt bad and could certainly understand his concern given that wild Indians were in the vicinity.

We had completed all the work we could and began to plan our return to Cochabamba. We visited the YPFB offices in Camiri and discussed our results with their exploration manager since it was necessary for us to keep YPFB informed. Just before leaving Mandeapequa a courier arrived from Camiri with tragic news.

Warren's two-year old son had arrived with his mother in Cochabamba just the week before, had fallen into the small ornamental pool at their new home, and drowned. She had just leased the house and moved in when this terrible accident occurred. The Company wanted Warren to return immediately and a stunned Warren left with the driver to Camiri. Unfortunately the weather prevented his flight back for several days before he could join his wife and fly back to the States.

I was eager to be with the family that would be arriving very soon and meet the other returning field party as well as the newly arrived personnel with whom I would be working in the coming years. Apart from the shock of Warren's baby son, I felt comfortable in this new job in a foreign country and with the local people who were friendly. Jacques, our party chief, would be returning to Venezuela at the end of this field season. He had been helpful and patient with a green crew as well as very considerate of our Bolivian personnel and the local people. He had been a great companion and I hated to see him leave.

6. Libby: Lima to Cochabamba with Frank

After that strange episode in Lima that night, I couldn't get back to sleep again in Dot's house. I could see the skies were lightening and there was too much to do and think about. I dressed, packed up everything and then the boys were awake. Breakfast was typical South American style although I didn't know it at the time, *cafe con leche, pan, y un poco fruita*. I was still a bit shell-shocked when the company representative from the Peruvian office arrived to take us to the airport. I bid Dot, who was still half asleep and a bit stunned herself, a hasty farewell and we were off to the airport in his rickety car.

Fortunately, we were delivered into the hands of the new personnel manager for Gulf's Bolivian operation, Frank Dozier, an American, who was not only fluent in Spanish, but spoke English better than I speak English today. He had been in the Venezuela office with Gulf for some years, had married a Venezuelan of Spanish descent and she and their two children would soon be arriving in Cochabamba. Frank took over completely: the passports, tickets, found our enormous amount of luggage that had been left at the airport, and even managed to keep John, who had decided the airport was a race track, under control. Frank would just point at someone and I can only imagine that he would tell them to 'grab that little boy!' since that is what they would do.

We flew at altitudes over 20,000 feet in a Bolivian plane that was not pressurized and Frank and I kept the hissing oxygen tubes in our mouths while we held the tubes to the boys noses. Thankfully they soon fell asleep and stopped fighting us. The smell of vomit permeated the cabin as we flew over famed Lake Titicaca and those enormous volcanoes with their gaping mouths. Everything seemed painted in a gray-brown wash. Moon stuff. These barren mountains, new mountains as the geologists would say, were so unlike our green forested Rockies.

When we descended to the La Paz airport, Frank left the plane to meet someone in the airport for the hour that we were scheduled to wait for new passengers. I could see nothing more than a few flat buildings in the distance, against a panorama of those bare mountains, with just a little snow this November springtime below the equator. While the boys continued to sleep, I sat in that 12,000 foot altitude looking out the window, probably hallucinating, because I could have sworn I saw Llamas floating by not far from the plane.

When Frank returned and we were in flight to Cochabamba, which was an hour down the mountains at a livable 8,500 feet, we were finally able to discard the oxygen tubes. Frank and Pilar were soon to become great friends of ours and we began to fill in the spaces about our families and each other. I asked Frank why he had been in Lima since it was highly unlikely that the company would be sending him to meet and escort all the incoming wives from Lima to Cochabamba. The father of two very young ones himself, he began to unburden himself as to why. He had come to accompany Warren, the geologist who had been with Eric in Mandeapequa, his wife, and the body of their two year old son, to help expedite them through customs in Peru and on their way to the States.

I was completely dismayed with this dreadful story as was Frank. We had looked forward to knowing these never-to-be friends who had arrived in this strange country and had to suffer such a terrible tragedy. The boys began to stir and then talk excitedly about their daddy. The mountains were now a friendly green and as we descended, it was wonderful to view that big, beautiful valley where we would be living. But a little voice inside me kept wondering if we had done the right thing, bringing these two innocents into this strange foreign country.

When we struggled out, Eric said we looked like three apparitions framed in the dark doorway of the plane. Admittedly there was a lot of inherited fair skin and blonde hair but the past 48 hours had distinctly added to our pale appearance. The boys fell into the strong arms of their tanned and slim father who now hardly looked his twenty eight years, for a tender reunion. I waited my turn. He soon piled us into his jeep and began to drive us around Cochabamba.

He first attempted to show us the plaza around which was one of the town's only paved roads. There were a few uncomfortable looking benches under scrawny palm trees where some old people sat, dressed in black. No response from us. Desperately trying to please and really not able to understand, with only my brief explanations of what the past forty-eight hours had held for us, he drove us next to the market place with its colorful Indians and powerful smells. Still no response. I think he expected some amazement, intelligent interest or just curiosity but we just wanted to go 'home', wherever that was, and right now.

7. Eric: New Office, Typhoid, and Hepatitis

When I arrived back in Cochabamba my first priority was to find a house for the family, and Frank, the personnel manager, offered to help. As one of the first to arrive, I was lucky to have found a two-story house with a yard surrounded by a six-foot wall in a good neighborhood close to town. A plus was that an American family with Point Four (a USAID program) lived across the street and had three young boys, the eldest a teenager. I knew I would be on field trips for extended periods, might be out of contact at times and petty burglary was common. There were only a few cars as yet in the small American community and I wanted Libby to be close to other Americans. The house came with Juan, the gardener, who would sleep nights on the grounds and that gave me comfort. I had done all I could and I was never happier in my life as when I saw Libby and the boys disembark from the Panagra flight, albeit pale and obviously weary.

Gulf was the first oil company to establish headquarters under the new petroleum law in Bolivia. We had a temporary exploration manager, Dave Kepple, the chief geologist from the office in Lima, Peru, who joined us with his wife, Barbara, and children, and introduced us and our families to life in South America. They brought with them a casual lifestyle with such imagination and humor that we all relaxed and felt that anything could be accomplished. More geologists had arrived or were arriving from the Cuban Gulf as well as two Swiss Geologists, a Canadian, a Dane with British citizenship and several American geologists, some with Gulf from the States as well as an American geophysicist. The families were just beginning to trickle in as their husbands could find housing.

When the general manager, Ben Hake, a long time Gulf employee and Geologist with experience in Mexico and Canada arrived, he initially spent most of his time in La Paz with the Bolivian company representative. They were working out the details for this new operation with laws in a new revolutionary government. Part of the agreement was that he had to commit the company to hire Bolivians with appropriate experience. His age, his experience, command of the Spanish language, as well as his authoritative manner, impressed the government officials and that helped our work immeasurably. Ben was essentially a trailblazer.

Gulf needed to bring in all of the equipment for exploration programs including geophysical contractors and their crews and the drilling equipment with their personnel. The aerial photography had already begun. Later we would hire our own flight support company and bring in airplanes and helicopters. This was complicated by the fact that Bolivia was a land-locked country and all heavy equipment that couldn't be flown in had to come into the port of Arica, Chili, or from Argentina, and be transported inland.

The permanent office was shaping up. When I left Bolivia three and a half years later, the exploration office staff had grown to about eighty people and approximately seventy percent were Bolivians. Within those years Paul Truit became the exploration manager, Roger Heggblom was the geophysical supervisor and I became the geological supervisor. Our first experienced Bolivian geologist, Eduardo Rodriguez, who was married with children, would accompany me on several trips in the coming year.

Now we were in a new building of several stories near the central plaza. The offices occupied the first two floors and there was an apartment on the top floor where Ben and his wife, Kirby, lived. Soon after we had established our offices, a graffiti hammer and cycle appeared with the scrawled words *Afuera* (get out) Gulf Oil, signed with the initials PCB, (Communist Party of Bolivia). This was a daily reminder of the fragility of the Bolivian Government where Marxism wanted a foothold in South America. Another almost daily reminder of their presence was fresh human excrement left at our office entrance.

Get out Gulf, the oil is ours.

After returning from Mandeapequa and writing up our field reports, Jacques returned to Venezuela. I made a short reconnaissance into the mountains with a small crew and somewhere on that trip I drank some lousy water or ate some bad food and I came back to Cochabamba with typhoid fever, despite my shots. Christmas was a blur of fever and before I fully recovered I turned yellow with hepatitis. Fortunately, Danny Juarez, a Mexican American from Los Angeles, in charge of the growing commissary, was able to find the right medicine in La Paz for the typhoid but the hepatitis took six weeks of bed rest. Everybody else in the house was healthy but we soon discovered that hepatitis was making the rounds of the community. This type of hepatitis was carried through food and water and at this time there were no known medications or vaccinations. The only medicine prescribed by the local doctors was a product called Alcibol, an awful tasting liquid that was supposed to help the liver.

From the time I had left Shreveport and returned to the bush on my next field trip I had lost 40 pounds and could get into my Marine Corps pants again. The boys were such a joy but a trial at the same time, especially while I was in bed. Having never seen me sick before, they really couldn't understand why I just wouldn't get up. They wanted to jump in bed with me for a rowdy time every morning. I had been living in such quiet that even the city noise with it's clatter, rumble and constant car horns kept me on edge. I was used to the quiet of the *selva* with only the occasional sounds of shrieking birds, monkey jabber or muttered Spanish by our guides. Our four-year old John was extremely active all his life and Paul tried to follow in his footsteps. Libby had her hands full. I was very thankful that she was settled in our new home, that we had a good cook and other servants to help her with all the responsibilities.

8. Libby: Home?

As we drove towards our future home, first impressions were pretty grim. No trees, grass or shrubs were to be seen in our residential area, only dusty, unpaved roads with deep pot holes, all under a merciless sun. We passed a woman with those full, colorful skirts and tall, mother goose white hat, scrubbing clothes while squatting by a tiny stream running down the wide rocky road. The only green growth to be seen was behind dun-colored adobe walls that were cemented on top with broken glass shards meant to deter intruders and stray dogs.

We arrived at a black metal gate that was opened for us and driving in, we were in another world of lush grass, many beautiful gardenia and rose bushes as well as huge geraniums climbing to the top of the wall. Our house came with a gardener, Juan, who would keep everything beautiful and flowering until the night of Good Friday. That night Jesus was supposedly in his tomb and couldn't see, so thieving was acceptable and every blossom just disappeared from gardens and reappeared in the local market to sell for Easter.

The house belonged to an exiled politician who had fled to Argentina but his wife was there at the house with a translator. As we walked through, she impressed upon us the value of their possessions and advised us that salaries for servants must not exceed a certain amount, anywhere from three to five dollars a month (!) depending on the job. Clearly these people were to be our slaves and the locals would be very displeased if we *Norte Americanos* upset the social and economic system. Despite travel exhaustion and the fear I had felt when I saw the ornamental pool in the garden, I was delighted when I saw the exotic furnishings and dark woods, but then we arrived in the kitchen and I noticed there was only one faucet. It was explained to me that if there was hot water in the kitchen the servants would only waste it!

The next day servants began to arrive, recommended by office staff. The cook, Aurella, appeared to be totally in charge. There was also a laundress, Dominga, a young boy to run errands and Rosa, who would look after our boys. According to Aurella, my place was to sit in the living room and knit, have tea with friends, discuss menus, and occasionally shop. A Bolivian teenager next door, who wanted to be a pilot and loved any opportunity to speak English, would help me with translation from time to time. About all the direction in Spanish that I could give at this point was about water. *NO AGUA* in the ornamental pool, *agua caliente* for washing dishes and *agua filtrado* to drink. But getting them to do it without watching every move was something else.

Having been raised in the South, the boys knew how to drink water from the faucet outside and although we had emphatically warned them, they soon became quite ill with 'Inca's Revenge' as we called it. It did, however, teach them a much needed lesson that they would never forget. We kept our bathtub full of water for flushing the toilet in daytime and the brown sediment at the bottom impressed John. Paul didn't like his stiff beige diapers either and that helped to train him sooner than I had anticipated. It's positively amazing how quickly children can adopt to new ways as well as learn a new language. The hapless maid, Rosa, was run ragged by Aurella and I soon found myself in charge of my rambunctious twosome, which suited me fine.

Aurella was a typical Quechua Indian about twenty years my senior with a long, strong nose and a large, powerful body. The servants quarters were surrounded by a cement patio behind the house and she would emerge each morning brushing her long, jet black hair and smooth it back to braid while watching Dominga beat and scrub our clothes with a soap that made no suds. We only had water for washing, flushing and bathing in the early morning before the town awakened and business began.

When Dominga would begin to spread the clothes on the cement to dry, Aurella would adjust her many skirts in studied movements and then make her way into the house to 'take over'. She seemed very clean and was a good cook and for that I was grateful, but when I was told in her most imperious manner that *no senora de la casa* could answer the telephone, I began thinking 'replacement'. I watched Aurella and learned how the daily enormous cauldron of soup for the servants was cooked on the kerosene stove with bones, vegetables, rice and the red hot peppers. Then I told Eric on one of his trips home that I wanted to fire Aurella and would he please do so because her controlling and authoritarian manner intimidated me.

He wasn't at all sympathetic to my reasoning and tried to convince me otherwise. After all, she served him coffee and hot biscuits as soon as he came downstairs in the morning, hovered around him and virtually treated him like a king when he was home. He was *El Caballero*, the master supreme who could do no wrong. Besides, it WAS true that although improving, my Spanish was still very limited and to care for the boys and manage a household in a Spanish speaking country was presumptuous. So I conceded defeat but immediately set about to turn my minuses into pluses.

When I knew I could trust Rosa with the boys, I began to venture out in those coughing taxis with the springs poking out, into the town and to the market. I attempted speaking Spanish everywhere I went in town and the locals always encouraged my efforts, especially since they only knew a few English words themselves. Our Spanish speaking friends say that Spanish (the language was called Castellano in Bolivia as in Spain but I shall call it Spanish) is much easier to learn and speak than English, which is true, but depending on any new language for your survival is tough stuff. It was a great help when I learned and put into practice the fact that the vowels in Spanish are always pronounced the same. What a difference from English!

Downtown vendor spinning wool.

I decided that Spanish would get all my effort although Quechua also surrounded us, especially in the marketplace. It was positively amazing how quickly our John picked up not only Spanish but Quechua. Once I was searching for the word 'pork' in the kitchen making noises like a pig to Aurella and John tugged at my skirt and said, *cerdo*. The only time I was ever ridiculed speaking Spanish was by several German store owners who would only speak their broken English with me and thought that the Spanish language was beneath a tall, fair woman. It didn't take me long to figure out which side of WWII they had championed.

It soon became obvious that the Company couldn't run a business with only one or two Bolivian chauffeurs and so at the Company's request the local government set up an evening meeting for all of us to take a drivers test together in a small school room. The Bolivian uniformed officials nailed a large map of Cochabamba on the wall, allowed us to study it for one half hour, took it down and then led us individually into another room where they proceeded to point out roads on maps for us to name until they found one we couldn't. If anyone passed the test, I don't remember. Our temporary manager, Dave, had been determined that we wouldn't pay bribes but no kind of determination could survive theirs. So the Company had to make other arrangements and I was a special case. While shopping in New York, my wallet was stolen but with no charge cards in those days, it really wasn't a problem except that I also lost my driver's license. Now they said I must demonstrate my ability to drive a car.

One Saturday morning, Eric, Frank and another geologist in town accompanied me to the site of the testing. I was to drive a Company Jeep, which I had never driven before but was assured it was not a problem. The location was surrounded with those dusty, dented, half dismantled taxis and except for me, prospective taxi drivers were the only ones to be tested. They all laughed and pointed at me in my yellow dress, that I would dare such a venture. We were required to drive our cars backwards over a rocky, running stream and then back up a mountain between rows of adobe homes without stalling. Most of the would-be taxi drivers failed and despite a fair performance, I also stalled, to my humiliation, but to the enjoyment of the other participants.

Driving at night was a frustrating experience since you were not supposed to turn on the car lights until you approached another car, which you couldn't see but were supposed to intuit. Also you must honk your horn at every intersection, the location of which had to be memorized. On the way to a party one night with Eric, we forgot a gift of food that we wanted to bring and had to return to the house. We drove up in the dark, turned the lights onto the driveway and caught Aurella, Rosa and the errand boy passing all kinds of food over the gate to their menfolk. The next day Eric fired all three, much to my satisfaction, although I made a successful effort not to gloat. When

Eric kindly took Aurella home the next day and helped her unload her baggage, he saw her house was a veritable store of toasters, irons, radios and other appliances that had obviously been appropriated from elsewhere. We were grateful to have made our discovery before our possessions finally arrived from the States.

9. Eric: Asuncion de Guaryos

When I began to recover I was able to discuss the regional geology with Paul T and Ben. Paul T found a contact who knew people in the area that we wanted to explore. It was northeast of Santa Cruz along the edge of the Brazilian Shield, a region of unprospective basement rock. This area was without roads, part jungle and part savanna which could be flooded during the rainy season. There was one airport near the edge of the area of interest at Asuncion but the only real access was by oxcart, horses or by foot. It was decided that I would go with an experienced Peruvian geologist, Amancio Caldas, from our exploration office in Lima, Peru, to a ranch with a landing strip near where we wanted to begin our overland trip.

We flew into that landing strip from Santa Cruz to the Cochamanides Ranch which was on the Rio San Miguel, a tributary of the Amazon River, and owned by a family of Greek immigrants that raised cattle. We stayed at the ranch two nights and I remember watching how they made beef jerky, the delicious taste of their sweet pink grapefruit and unforgettable images of the locals who appeared from nowhere to collect their pittance and a ration of alcohol for their corn harvest. The rancher furnished us with two outboard motor powered dugouts and three people to be our crew and guides. These men would carry us and our equipment down the river to another ranch at a place called Quebrada Blanca, where the river intersected a trail. On the trip down the river we were looking for outcrops of sedimentary rocks which would help us define the eastern margin of the basin. The geology of the area was unknown and any surface maps available showed no details.

There were several Indian tribes that could be dangerous in

this area so we carried firearms and camped on the banks of the river during our four days of travel, keeping close guard. We saw many capybara, the largest known rodent about the size of a pig, as well as tapir. We saw no game that we could eat but one of the men caught piranha fish which was made into a good fish chowder but, however, discouraged our bathing. Fortunately we saw no signs of Indians, left our guides with a good bonus and looked forward to the overland part of our trip.

We stayed two nights in the ranch at Quebrada Blanca, also made with the white plastered buildings like all the ranches we had encountered. At night we dined with the rancher and his family who were delighted with foreign visitors in this remote area. We negotiated with him to provide us with several mules which were the favored method of travel in the *selva*. The next day the rancher insisted we have a practice ride with our mules. He asked me to bring my rifle and when we saw a rhea, an ostrich like bird, he asked me to shoot it. When I asked if we would be eating it for dinner, he laughed and said no, that he just wanted it for the tail feathers to give to his wife and so, I had to oblige him. The morning we left we were served an enormous breakfast, which included choice cuts of a young beef, slaughtered for the occasion. The unexpected 'treat' was the delicious sautéed ovaries.

Our party consisted of Amancio and myself, one guide and four mules, including a pack mule that carried our food and equipment. By this time Amancio and I were getting well acquainted. He proved to be not only my translator but became a friend and introduced me into how one could live off the land in this kind of country. The next several weeks we searched for diagnostic outcrops that would help explain the geology of this unmapped area. We were fortunate enough to find enough to define the boundary between the sedimentary rocks to the west and the basement rocks to the east.

During this trip to Asuncion, which had the only airport near our destination, I learned about the Spanish colonization of central South America, besides geology. One night we made camp at the site of an old colonial mission church now abandoned at Yotau that was one of the furthest outposts of the Jesuit Missionaries who had provided conversion as well as European education to the local people. Although this church was crumbling after two hundred years, the roof timbers were still held together by a certain kind of unbelievably durable vine.

When we arrived in Asuncion, a town of several hundred people, we went to the local mission to introduce ourselves. We wanted to explain that we were not Protestant Missionaries, competing for souls with the local Catholic 'padres', but scientists studying the geology and we needed help in finding guides. To our surprise the Father Superior was an American. For me, after weeks of struggling with Spanish, it was a great relief to meet this good Franciscan Father from Arizona. He was also delighted with our company and opened an old and very small refrigerator in his office that was full of Cochabamba beer. He also offered us local cigars, tightly rolled cheroots, that were made by local Indian women who rolled them on their thighs. They were powerfully strong and had a few wormholes but the beer was a real treat and the company good.

We wanted to examine an outcrop that had been reported by air travelers. The people who were most knowledgeable about that part of the *selva* were elderly, afraid of wild Indians, and wouldn't accompany us. We finally found a young school teacher who was also a hunter, had seen this hill that resembled a geological remnant and willing to take us there. Eventually we also found two more hunters to carry our gear, although this trip should take us no more than two days. We hired a boat to take us across the Rio San Miguel, had a quick uneventful trip but identified the outcrop to our satisfaction. I was awakened in the morning, after we had slept at the foot of the outcrop, to a very loud sound like a railroad train close by. The locals laughed at my amazement while explaining that it was a troop of howler monkeys.

We returned to fly out of Asuncion to Santa Cruz and then Cochabamba. After a few days Amancio flew out to Lima to be with his family. I had looked forward to a long friendship with Amancio but

I never saw him again. He was killed a few months later in an airplane crash that was carrying him into an area that he was studying in eastern Peru. Ironically, the only map available to search for his plane was one that he drawn up a few years before.

10. Libby: Augustina and Hepatitis

Cochabamba was a picturesque city in a large mountain valley and looked a little like the pictures of southern Spain. Situated in this dusty elongated bowl, and surrounded by green mountains, it had a beautiful climate where you could wear a sweater in the evening, have a fire in the grate at night and a blanket on your bed almost year around. The days were bright and clear, we seldom had rain and sometimes a skiff of snow. Located just above the jungle, it also had enough humidity to make the climate truly enviable. But there the local amenities ended.

Bolivia was the poorest country in South America, in the middle of a severe inflation, the victim of corruption and power struggles with violent revolutions constantly brewing. Rich in mineral wealth, it was the world's largest producer of silver in the Sixteenth Century, followed later by tin, and now, exports of natural gas. The Quechua and Aymara Indians had been virtual slaves of the predominately old Spanish stock since the time of the Conquistadores. The democratically elected Socialist Government in 1952 nationalized the silver, tin, copper and gold mines and was supported by the USA State Department with Point Four programs. They began to develop new industry, build roads and bring military aid. We often depended on the military, they became our friends, and some still are until this day.

The population of Cochabamba when we arrived in 1956 was about 100,000, consisting of about 53% Indians, 32% 'mestizos', and the remainder an amazing polyglot of Spanish descendants, Egyptian as well as European Jews (in the year 1938, Bolivia was one of only two or three countries in the world accepting Jews fleeing from Hitler), escaped Nazis, and other refugees from World War II as well as East Europeans of all backgrounds fleeing the expanding world of Communism. Employees of other oil companies began to filter in, Americans, Canadians, English, Swiss, Dutch, etc., but we made up only a very small part of the community. However, within the first year of our arrival, we *estranjeros* had established an English speaking school as well as the business of oil exploration.

Poverty in Cochabamba.

Downtown the Germans had sausage stores, the Viennese had coffee shops, and the Yugoslavs sold pungent herbs, spices and staples, all at ridiculously high prices. The Bolivians of Spanish descent ran the pharmacies stocked with unidentifiable products and the stores that displayed dried meat hanging in the windows. Inside, cans of local tomatoes and a gray looking strawberry jam were sold along with a miserable kind of hard, dry parmesan cheese and canned beef imported from Argentina. They also had light bulbs that lasted a week, soap with no suds, shoe polish that wouldn't polish and shampoo that really couldn't shampoo.

The Indians sold some food and grains on the sidewalks downtown but one needed to be careful where one walked. Those big skirts gave women good cover to squat and do any bathroom activity, anywhere, and the men urinated on any wall at any time or place with no shame. The dirty sidewalks were depressing but the downtown smells never reached the same peak as in the open market. Thank heavens it was open.

Fortunately we still had Aurella and her crew during the period Eric was so ill. Now I was delighted to be alone without Aurella but I knew I couldn't survive for long with only a laundress and gardener. Although the telephones seldom functioned, I could generally find help through the office. Other wives were as busy as I, trying to sort out their lives. I soon found Corina but she had family, couldn't 'live in' and was better at cooking than shopping.

One night we were awakened at 3:00 AM by several inebriated bachelors who came by with the Police Band to serenade Eric on his birthday! Then they expected breakfast. Fortunately, I had recently bought some eggs from a house vender (we had to put them in a bowl of water to assure their freshness and if they floated, you passed), we had some coffee and I had made some bread just that night. This was no ordinary life style and I had to adjust.

Juan disappeared for a few days and when he returned I questioned him with the help of my teenager next door. I discovered to my dismay that he had to bury a child who died of malnutrition. It was my first 'moment of truth' and I had many in the fifteen years we lived in foreign countries. I began to understand the great poverty the Indians were enduring and would have liked to have hired several and paid higher salaries to ease my conscience. However, I just knew I couldn't live with a house full of servants and the Bolivians were already giving us trouble for paying *demasiado* (too much).

The biggest chore was buying food in the open market and I had to shop with bills of 10,000 Bolivianos that only had the value of a dollar. Good vegetables and fruit were scarce to nonexistent and the open meat market was a sight to behold, with vultures overhead peering down with long red necks and excited with the smell of blood.

It was necessary to boil the drinking water and milk for thirty minutes, which pretty much ruined the milk, and then the water had to be filtered. All the fresh fruit and vegetables, unless peeled or cooked, had to be soaked in a water and Clorox bath for fifteen minutes. There was no raised bread to be bought and since the electric oven only functioned at night when electricity was no longer in demand by businesses, the dough had to be risen and ready to bake after 10:00 PM. It was all too much for me.

I finally found the most wonderful combination cook/maid/nanny ever, Augustina, who made the rest of our stay a happy experience. Just my age, she was a Mestizo, wore western style clothes and came with her little daughter, Inez, who was born the same day as our Juanito, and the two immediately became friends. Although Augustina could be shy at times, she easily assumed authority, was honest, totally dependable and exuded confidence in her performance.

One morning, while Eric was at home, I was too weak to get out of bed, and in one day, I turned yellow from head to toe and could hardly lift my little finger. I had been so careful with water and food and couldn't understand how I could have gotten hepatitis. However, it turned out that it can also be sexually transmitted. A German nurse, Anna Maria Grundner, whom Ben had hired, came daily to give me Vitamin B shots and instructed Augustina to feed me peeled grapes, that, as explained by her, would get sugar into my liver and give me energy. Eric had to leave on a trip but Augustina ran everything in the house and cared for the boys and me. One day Ben and Kirby visited me while still in bed and Ben asked if I wanted Eric to return to Cochabamba. I asked if Eric would have to travel back to the same place in the *selva* again, which had already taken some time, and Ben said "yes". So I thanked him but said, "no".

It wasn't much longer before our possessions, toys, records and player, Hallicrafter radio, odd pieces of furniture and general household accumulations arrived and we were one happy family to connect with our personnel belongings. I had been wondering if I was pregnant before I became ill and yes, I was expecting and now we were terrified that this baby might suffer some ill effects from my hepatitis. I was so careful the next seven months you would have thought we were running a hospital inside a nunnery.

I would have gone home for the delivery but the separation from Eric would be at least six months and I would have the boys with me and be staying with various families. The boys were so happy in our new life and had never had any serious illnesses during our travels. Perhaps because they had already suffered scarlet fever, mumps, chicken pox and measles, in that order, that they contracted one after the other from the doctors office in Louisiana. Also this was my third, I'd never had any difficulties before, and after all, my oldest sister had given birth to her first child, now in her teens, in Karachi, India!

11. Eric: Missionaries and Ayoreo Indians

During the remainder of my second field season I made several more trips to continue our reconnaissance. We wanted to find the contact between the older rocks of the Brazilian Shield and the younger sedimentary rocks of the Chaco basin to the South. We believed these rocks were of an age equivalent to the oil productive section in the Andean foothills two hundred miles to the west. This could open a new oil producing area. I was accompanied on these trips by Eduardo, who had experience studying these types of rocks in the Andes. He spoke some English, taught me to speak much better Spanish and expanded my knowledge of Bolivian customs and politics.

Ayoreo hunter gatherers at missionary air strip.

Several airstrips had been built in these isolated areas by Protestant missionaries who were most helpful in our exploration efforts. Several times we landed on these airstrips when the wild Indians had come into the mission to trade. We saw few Indian women and they were dressed in shifts, obviously furnished by the missionaries. The men were often nude or wore feathered ornaments and generally carried weapons, spears five to six feet long as well as bows and arrows. One end of the spear was blunt and wedge shaped for digging roots. A Shaman from one group wore a jaguar head dress that he traded with me.

These nomadic hunter gatherers survived on roots, tortoise and small game, and were expert at harvesting honey from the wild bees, just as our ancestors did. They cut railroad cables to make tips for their arrows that could penetrate a car door, as experienced by Andy, one of our geologists. Although some of these Indians were wild and dangerous, others had obviously mixed with Mestizos and Europeans. We were told that the Indians and locals raided each other to capture children to raise as slaves. It was documented that five Missionaries had been killed in this area five years before by the Ayoreo Indians. These were the same people with whom we had met and traded with at one of the mission camps. During my stay in Bolivia a missionary who was visiting a drilling operation in northern Paraguay, in which our Company was a participant, was killed by Indians not far from the camp. Consequently, the Bolivian and Paraguayan governments usually sent military escorts to accompany us into this region. Another factor involved was that Paraguay had won a recent boundary war with Bolivia and tensions were still high between the two countries.

On subsequent trips into the *selva* we carried combs, mirrors and beads to leave as gifts at their crude shelters, hoping they would think we were friendly. By the 1980s, most of these peoples had been decimated while very few had been acculturated. This was the consequence of the missionaries, disease, the settlers with their development of farms and railroads, as well as the oil and gas industry. The cost of 'progress'.

On the southern margin of the Brazilian Shield, the Brazilian Bolivian Railroad Commission was building a railroad from Corumba, Brazil to Santa Cruz, a railroad that had been under construction for "too many years", according to Eduardo. The Bolivians had holidays, the Brazilians had holidays, some political, others religious, plus union problems and revolutions in both countries and there were refugees crossing the border periodically. We were able to use rail travel to carry us to several of the main towns to study and measure outcrop sections. When trains were not scheduled or available, we sometimes borrowed handcars which the two of us managed to work together. We were anxious at times, having heard a story about the Indians chasing two railroad workers into a small maintenance shed where they locked themselves in. Finding railroad flares inside they managed to shoot one off out the window for help. They were surprised to hear the Indians began to shout "Viva St. John" and then leave. These Indians were familiar with the legend that the Jesuits of long ago celebrated the feast of St John with fireworks.

In Robore we met a friend of Eduardo's, Colonel Quintanilla, who was in charge at the principal military outpost in eastern Bolivia. One afternoon we were invited to a military 'beerfest' with the officers and they entertained us with stories of their encounters with the Indians as well as political refugees that had taken up residence nearby. Years later I learned that the Colonel had been appointed Bolivian Ambassador to Germany in the sixties, where, for some reason I never clarified, had been assassinated. However, there had been two failed military coups in Bolivia during that time.

In nearby Santiago, where we stayed for at least a week, we found some interesting geology. After introducing ourselves to the *Alcalde*, we went to the church that had been an old Jesuit Mission, and now maintained by an order of Italian nuns. The Mother Superior entertained us with tea and cookies and a story about an Ayoreo young girl who had been left to die in the *selva* because of illness, was found by a hunter, and brought to the Nuns to raise. She was an extraordinary child who quickly learned Spanish, Italian and French, taught them about local herbs and plants and was a musical prodigy, learning both the violin and piano. She was so exceptional that they sent her to Rome when she was a teenager but she died in Rome while in her early twenties from some unknown ailment.

We found quarters in the back of a small restaurant with a diesel generator nearby that worked intermittently. We were awakened early in the morning by the braying of mules and after breakfast we would take our prepared lunch and head for the abundant and interesting outcrops with our helper. We would return with sacks of rock samples and write our notes before it was dark.

One evening a week the locals would fire up the generator to show a forgettable movie outdoors on the wall of the largest building in town. Another amusement was watching the Venus flytraps, or carnivorous plants, that lined the plaza, catch and consume the bugs. Just before leaving Santiago, Eduardo spent the evening chatting with locals and returned with a big smile. Apparently they were convinced that our *petrolero* story was phony and that we were really looking for Jesuit gold. In 1765, the King of Spain ordered the Jesuit fathers, who were interfering with his plans to colonize this area, to close their mission and leave immediately. With not enough time to carry out their treasure, the local legend had it that they buried a large quantity of gold nearby. I remembered that the Franciscan Father in Asencion told Amancio and me that besides the cigars, the local Indian women would also bring small amounts of gold dust. Without Eduardo, my sensitive, self-effacing yet influential Bolivian companion, few of these historical stories would have reached our office personnel.

My last field trip was a short reconnaissance with Luis, Victor and 'Loro' (parrot) who was nicknamed for his hooked nose. I wanted to return to the eastern Parapeti block to see some outcrops at the corner of Paraguay at Cerro Ustares. We stayed no longer than one night after obtaining our samples and returned on the overgrown track with some apprehension. We had seen signs that the wild Indians were in the area. I was riding in the Dodge power-wagon with Luis and we thought we ran over a log but looking back I saw a huge snake slowly moving into the underbrush. It was an injured ten foot boa constrictor and we decided we had to end its misery. Although I had expected many, this was my only real snake experience in Bolivia and I was relieved that we never had a confrontation with the Indians.

By late 1958 the reconnaissance geology had mostly been completed and more detailed studies had begun. The drilling rigs were arriving and we needed to select locations and confirm with the geophysical crew and their seismic data. At that time I was mostly working in the Santa Cruz area with several American and Swiss geologists who had recently arrived. We followed up and confirmed some structural leads with the help of aerial photos and one surface prospect that seemed very promising. I was then asked to supervise the evaluation of the exploration wells which meant that I would be spending most of my time in the office. This suited me well since we were now five! Libby had given birth to another wonderful son, Mark, a Cochabambino, in February 1958.

There had been an established English Speaking Club in Cochabamba for some time. It was a mixture of Europeans, refugees and locals. Now we were bringing in many new members from the oil business and since I was now a more permanent resident, I was somehow elected President. We had a weekly bridge night and one incident remains with me. Two German women who were partners often spoke in German and one night they played with one of our military personnel who was a multi-lingual Hungarian. He told me with some amusement that they were cheating. We then decided that all club members would henceforth speak only English in the English Speaking Club. Their bridge game fell off after that.

12. Libby: Move and Mark

When our manager, Ben, who had come to symbolize dignity and paternal authority in the community, gave a speech and our flag was raised at a Fourth of July ceremony, there was hardly a dry eye among us. Most of us had given little thought to our country since World War II but life in Bolivia had opened our eyes to the great advantages that we had in the USA.

Still, our government did foolish things. Point Four had just built a dried milk factory producing a product called KLIM (milk spelled backwards) which the Quechua hated, preferring their own milk, be it contaminated with brucellosis or not. The USA shipped cheddar cheese to Bolivia in seven pound cans, excess produce from US farmers and stamped NOT FOR SALE and we were explicitly told not to buy it. Corina had given me a portion of that cheese which had routinely been distributed to her family by the church but no one in her family liked it and she couldn't even give it away to her friends. That's when I decided to buy the cans when I was in the market, figuring the venders could take the money and buy what they wanted. How we loved it! Point Four also built a million dollar plant just to put the red in the strawberry jam.

At the same time the clever Marxist Bolivians were changing the markings on the much desired bags of grain from US To USSR and the Russians were broadcasting in Quechua on the short-wave radio stations. A visiting state department representative held a meeting for us wives and attempted to convince us that they were doing what was right for our country. However, we came away feeling that the farmers were just being subsidized, and our government was stupid to produce products in a country, as well as donate products, that no one wanted.

Edible food was always on our mind. A farmer from England, whose wife had invited several of us for tea and bridge, offered us a chance to buy cuts of pigs that he was just going to slaughter. How we jumped at the chance! Sadly, he returned before we left to tell us they were full of trichinosis. Whenever Eric was in town, we would often take our John with us to market and buy a hand full of peas here, carrots there. The waste of food, water, and paper products in the US plagues me to this day.

We always attempted to bargain with the Quechua since that was considered the way to gain their respect. They spoke mostly their own language and so early on we learned to communicate with gestures. Unless Juanito could translate. The women did most of the vending as well as the work and were naturally happy to embrace the Communist idea of equality with the men. Coca leaves and a sort of cake made with lime were sold in the market and commonly chewed by the men. Many of the men were not much more than beasts of burden, carrying supplies over those high mountains and the coca leaves enabled them to continue for days in the high altitudes with little food. Although Che Guevarra had not yet arrived in Bolivia by this time, the stage was being set for his arrival and eventual demise. Unfortunately, in some areas the Quechua thought that the mapping work done by geologists was an effort to take back the land that had been redistributed with the 1952 revolution, and the Communist party capitalized on this misinformation and were constantly agitating.

Eating out was very rare since it made us sick so we entertained at home. We were soon called upon to help entertain stateside business visitors with cocktails or even dinner with little notice. Alcohol, mostly in the form of hard liquor, but also wine and liqueurs, were mostly introduced to us by the Europeans. Beer or various alcoholic punches were essentially all that Eric and I knew. Combined with our youth, the challenges we were all facing and inexperience with alcohol led to some unforgettable parties. One of the Swiss geologists, Guy Chamot, had his wife bring in an enormous round of cheese along with Kirsch, which they kept under lock and key. It was truly a prize invitation to share cheese fondue at their house.

There was no way to avoid shopping in that very dirty, smelly market place. I was stared at, bumped into, ignored, laughed at, constantly cheated and heaven forbid that I would have the nerve to take a picture. I would buy the whole *lomo* (filet mignon), bring it home, scrub off the fly droppings and then because it was so tough, it would take both Augustina and me to push it through the meat grinder. Range-fed beef has a completely different taste than corn-fed and Bolivian range-fed tastes different from any other range-fed, but with enough chopped onions and tomato sauce, it was edible.

I brought home a large sack of flour one day in a taxi and Augustina found it full of weevils. I knew I would have no luck returning it and was ready to begin sifting them out but she said we must take it back immediately. Sweet, mild mannered Augustina could give quite a commanding performance when dealing with her own, and she was not afraid of the male sex. After a lot of Spanish and Quechua thrown back and forth, we got another sack which she carefully examined before we left.

Thanksgiving and Christmas were approaching and Augustina and I bought the best looking turkey in the marketplace and brought it home, determined to fatten the creature. It absolutely refused all food no matter what we offered. We finally decided it must be dispatched before it turned into skin and bones and put into the big freezer I had recently bought from someone returning to the States. For some reason I never determined, Augustina refused to kill it, she who would do anything for me and the boys, and so I approached the older sons of our American neighbors across the street and with their mother's approval, the challenge was accepted.

The Bolivians had told me that the only way to guarantee tender meat was to pour *chicha*, a local corn beer, down the turkey's throat and wait until it relaxed. I followed their advice and when the wattles turned white and it laid its head on the chopping block, the boys had a go. After a few awkward tries and no results, I couldn't stand anymore and almost six months pregnant, I wielded the axe and with one well-guided and controlled blow, I gratefully, and I must say, proudly, released the poor thing from its drunken stupor. It stayed in the freezer until we moved and it was more tender than any other bird that we ate during our long stay in Bolivia.

After we realized that I was pregnant, we decided that we would not tell our parents to spare them the worry. We checked out the maternity clinic in town and were delighted to find that Doctor Morales could speak better English than we could Spanish. Renewal time for our rental contract was approaching and we decided for security and convenience to move into a very spacious upstairs apartment with three bedrooms, ample servants quarters with bathroom, and three large balconies, one adjoining American neighbors next door. We would also overlook the only paved *Avenida* in town that we already knew well since that was where the religious parades and colorful, lively dances took place during their carnival.

Our landlord, Branco Avanic, a friendly Yugoslav, and his warm hearted Bolivian wife and children lived downstairs. He had an aviary with beautiful exotic birds and the grounds were supervised by him. With great emotion, he would tell us everything we wanted to know about the Yugoslavian dictator Tito and the world he left behind, but only in the middle of his yard during the daytime and in a whisper, while constantly reviewing his surroundings to watch for any eavesdroppers. We were most impressed with his fear of local Communists.

In our new home, I blossomed literally and figuratively. The responsibilities of keeping a large house and garden with all its everyday problems eased. Not only had I learned what it feels like to be admired and despised at the same time but how to play parent to a very needy people. I had to let Juan go but kept Dominga, who would still wash for the family, and of course, Augustina and Inez. When the adjoining apartment upstairs was vacated, Frank and Pilar moved in with their two children, Consuelo and Michael, both of whom had become close friends with our 'Juanito' and 'Pablicito', who now spoke both English and Spanish interchangeably, and we had two and a half fun years as next door neighbors. Pilar helped guide me through and educate me in the ways of the Latin world for which I was forever grateful. About this time we added our wonderful little dachshund, Loompi, to our family, who travelled the world with us.

Our Mark was ten days late and Kirby decided a party in their home would hasten it along. I didn't do much dancing but had a wonderful evening and it worked. We went to the clinic in the middle of the night but Doctor Morales was on duty alone, and so when I demanded it, and demand it I did, Eric had the job of putting a gas mask over my face. At that altitude it was dangerous to give anesthetics, something I had not known before. I never dreamed that anesthetics were such a help at the end of delivery.

Despite Eric exclaiming Mark looked like a prune when he emerged and arousing our fears of the effects of hepatitis, he turned out to be brighter and healthier than normal, as were all our sons! He was born on George Washington's birthday and the American bachelors were at the clinic within two hours after his birth demanding that he be named 'George'. I had been convinced it was a girl since he hardly moved during the pregnancy, unlike his active brothers, not understanding until much later that it was the lack of oxygen at that altitude that had kept him quiet.

When we brought Mark home, after three days in the Clinic, the boys couldn't keep their hands off their new baby brother and Augustina thought he was her son. We had to put his crib legs in bowls of water to guard against the deadly *chagas* disease beetle, just in case, but otherwise his babyhood and childhood were quite normal. I returned with him to the doctor for a six weeks checkup and his wife was there also. He introduced me and when I presented 'Markicitos' to her, she smiled and responded in Spanish "Oh, I know him well." It seemed that despite all our precautions in the clinic, bringing in boiled water, linens from home and only Eric and I taking care of him, apparently when I was asleep and Eric gone, they would whisk him upstairs where they lived, for their dinner parties, so that all could marvel at his fairness, his veins that were so blue, and his downy white hair.

Markicitos, our Cochabambino.

13. Eric: Airplanes and a Volcano

While I had been working near Santa Cruz and the Parapeti block to the South, Paul T had organized a program for exploring the northern Chapare region which was the least populated and essentially without roads. There was no geological information on this territory and it became necessary to fly geologists in and out, mostly for day trips on helicopters. Subsequently and unfortunately, this area eventually turned out to be nonproductive. Our geophysical supervisor, Roger, who had arrived six months after me, had never worked in a foreign situation either, and had his hands full with several geophysical crews. The seismic crews were mainly focused on the Santa Cruz area with additional exploration in the YPFB prospects.

I was now scheduled to supervise the well site geology that was beginning for our first test on the Mandeapecua structure as well as some of the follow-up surface work in the Santa Cruz region. Roger and I reported to Paul T, who had worked for Gulf in Mozambique and Cuba and spoke fluent Portuguese and Spanish. The three of us had different backgrounds but our experiences complimented each other and we worked well together. Ben maintained a firm hand over the whole operation, which covered thousands of square miles and employed over a hundred people, expatriates and locals, besides contractors and their crews.

Now we had a big rig in Mandeapecua and a smaller rig which we used to drill stratigraphic exploratory holes. We were compelled to construct airstrips for the planes and helicopters where we could bring supplies for the geological parties. We contracted with an air service from Lafayette, Louisiana, that had experience with oil operations, and they furnished us with two fixed-wing airplanes and several helicopters plus their pilots and mechanics. They were excellent pilots and companions in this exciting place that we had found ourselves, and some of them eventually returned to the States with Bolivian wives. Periodically I made air trips to these drilling or field operations and two particular flights are burned in my memory.

The larger plane was an Aero-Commander and the other was a plane for short landing strips, a Helio-Courier. The pilot for the Helio-Courier claimed it was a difficult plane to fly and I discovered exactly that on my first trip in this plane. We were taking off from a short strip in the Chapare region and just 150 feet off the ground and clearing the trees when the motor quit. In the sudden silence, the pilot, Lynn Clough, and I just looked at each other. Several seconds later he managed to get the motor started up again. Neither of us could speak for several minutes and I was white knuckled holding onto my seat until we landed in Cochabamba. Shortly after that, the Louisiana contractor decided to replace the Helio-Courier with a Canadian bush plane, the Beaver, for which we were all very grateful.

However this was not my last encounter with the Helio-Courier. I was unaware that the contractor had sold the plane to a local air company and we scheduled trips with that company from time to time. We had a visitor from the Gulf Research and Development Lab outside Pittsburgh and he wanted to fly over some of the country where we were working. I decided, along with a Swiss geologist, Walter Hess, to take Mel Hill on a trip over the Chapare. My heart sank when I realized the Helio-Courier had been assigned to us but the Bolivian pilot assured us that he had been checked out to fly the plane. On his first attempt he couldn't get it off the airstrip. He taxied back and managed to take off but began to lose altitude over Cochabamba. I turned to the pilot and asked that he take us back. He argued that he could fly this plane over the 12,000 foot mountain peaks looming ahead but I demanded he return and we literally had a shouting match. He finally brought us down with such a rocky landing that he nearly broke the landing gear. That was the last trip on the Helio-Courier that anyone in the company made.

As we were the oil company with the largest operations, the government would sometimes request our help with geological or engineering problems. The Governor of the Province of Cochabamba was concerned with a community in the mountains on the railroad line about fifty miles towards La Paz and requested our help. The people in this village, Arque, were frightened about smoke coming out of fissures in the ground that had appeared after a small earthquake and feared that it might result in a new volcano. The people were panicky and threatening to abandon the village.

With both Ben and Paul T gone, the Personnel Department asked me to supervise a trip to determine what was happening. I took another geologist, Griff Lloyd, a Canadian, with me and we traveled to Arque by *autocaril*, a small railroad coach with a diesel engine. We were completely taken aback when we were met at the station with a brass band and most of the townspeople. Griff had recently arrived, spoke no Spanish and mine was just adequate at this time.

About 50 people plus the local *Alcalde* led us for a mile and a half to the valley where the smoke had been seen. There were vapors rising from several fissures and we weren't sure what this meant but it was obvious that we needed to have an opinion. We were the *Doctores*, the expert scientists and we should know. I did know enough about the regional geology at this time to realize we were several mountain ranges east of any active volcanoes in Peru. So Griff and I walked around, collected a few rock samples, bottled some of the fumes to be analyzed, and huddled together in English. Number one, we decided we must appear confident, number two, allay their fears, and finally, assure them we would let them know if there was any danger.

We returned to the village for a gay fiesta of tough chicken, potatoes and *chicha*. This was a mildly alcoholic drink that the women made by chewing corn and spitting it into an enormous pot and allowed to ferment with water. Neither of us was enthusiastic about drinking this brew which I felt obligated to describe to Griff, but I knew enough at this time to understand how political our position was in Bolivia.

The next day the headline said, "*Geologo Ericson Dice No Es Vulcan, Es Fumarol in Arqué*" (Geologist Ericson says it is not a volcano in Arque, just a smoking fissure). When Ben returned and heard the news and saw the headline, his only comment was: "The big new volcano in Mexico started the same way and if this one blows up, you had better be on the next plane out." I kept my eye on that area almost every day from our upstairs balcony while in Cochabamba, and so far, to this day, there have been no new active volcanos in that part of Bolivia.

14. Libby: Vacation and Fishing

When Mark was eight months old and our biannual, two month vacation time came around, we could hardly wait to return and share our experiences with family and friends. We traveled with Pilar, Frank and family and we arrived in Houston first. The next day both families went to the supermarket together and we just stood around, shocked at the size of bananas, apples, tomatoes and speechless as we looked at the rows of boxed cereals, cookies and crackers, shelves of paper products and refrigerators full of cold milk, cheese and wonderful looking meat behind spotless counters.

Neither John nor Consuelo, both just turning six, could imagine that these products were purchasable, edible and usable. They kept running down the aisles and pointing at things. It was almost too much for us adults to absorb, and for the children it was more grand than any Christmas of which they could ever dream. We parted from our friends and drove north and west to visit friends and relatives in Texas, Colorado, and Kansas. I had hoped to impress our friends and my many teen nieces and nephews with how fortunate they were to live in this country but few ever paid any attention. My mother only had eyes for the boys but my sisters, who had also lived in foreign countries, understood and listened.

While in Colorado we decided we must buy a little cabin in the mountains for future vacations but we never found time to look around. We continued our trip back east by train, stretching friends and relatives patience with our stopovers. While in New York, we attempted to fill everyone's requested purchases for Cochabamba, present and future gifts for Augustina, Inez, the boys, and our own needs. We were a couple of exhausted parents who settled down for the flight back.

We had mixed emotions about returning to the continuing battle to find edible food and safe drink, but the challenge of a new language, culture, experiences, 'virgin' territory for geologists and new worldly friends had hooked us but good.

Politically, some things began to change on our second two year stretch. When we had first arrived we saw the Communist logo painted on walls and heard of coups and of new presidents being installed but none of it really touched us. However, we were now being called at night on the telephone by disguised Spanish speaking voices that warned us to fill our bathtubs with water and avoid the office and the streets for several days at a time. We warned others and did as we were told. In 1958 Time Magazine wrote an article belittling Bolivia, suggesting that it be divided between its neighbors. This resulted in riots in both Cochabamba and La Paz with some killed, many injured, and the USIS office in La Paz being gutted by fire.

Another 'Ericson' in the community with Point Four was leaving after several years and went to a party given for him by some employees at a local tavern. A bar room brawl broke out and he was hit on the head and killed. For some reason Eric was required to identify his body. The Bolivians claimed he was CIA, that the party was a set up, and it might have been. His widow and children were given an apartment house of several stories in Miami by the government, or so we heard. My mother was beside herself and we sent a cablegram to assure her that Eric was okay but it was garbled on arrival. However she contacted the Kansas City Tribune and they finally clarified things.

One Saturday, Ben and Kirby took John and me, driven by their chauffeur, to a nearby village high in the mountains to buy some specialties at their Quechua market. I had never felt so uncomfortable before in my life. The Indians openly glared and turned their backs to us, would only give a price if asked by the chauffeur and would carry on an exchange of money and product in silence. The Communist movement was alive and well but I found it hard to blame the Quechua, considering how miserably they had been treated by their aggressors for centuries.

One long May Day weekend, despite it being winter, a group of us decided to travel up high into Indian country and find a river only accessible in April and May that was said to be teeming with big and tasty trout. The trout had been introduced by Germans who were involved with various business ventures as well as military instruction in Bolivia. German staff officers were advisors for both Bolivians and Paraguayan armies in the Chaco War before World War II. We drove with Paul T and Dorothy in our company jeep, there was a Scottish couple, Jack and Lydia Spinks, with Shell in their Land Rover, and two army men in their truck. We began to ascend the mountain road late Thursday night after a flat tire, and frankly were lucky not to have seen the narrow roads and sheer drops we traveled in the dark.

We arrived onto a high, flat plain called Viscachas about 1:00 AM and seeing lights and movement nearby that seemed to be the only habitation for miles around, we asked some Indians for permission to sleep in their stone hut. They removed their winter storage of potatoes so that five of us could stretch out in our sleeping bags on the dirt floor. It was freezing and I opted to spend the rest of the night with four men while Eric and the other two wives chose to sleep outside or in their cars. The Indians were celebrating the big Communist holiday, May Day, and played their reed flutes, and probably chewed coca leaves all night long in an adjoining room. I was awakened every half hour by chattering teeth or the snoring of the men, and in the background was the same strange, monotonous tune being played over and over again.

We awakened to find Eric covered with snow and to see those beautiful vicuñas wandering around the foothills of the snow covered mountains far in the distance. There was no vegetation, just scrub and rock, the atmosphere had no color and we felt like we were on the moon. We had been told that another group of fishermen were also on their way to Jatunrumi (pronounced Hot-tune-roomy) and about that time we heard them pass and so we were up and away, passed them, and after another two hours drive did manage to get to the stone huts first. The fishermen and women, everyone but me, downed coffee, scorned breakfast, grabbed some cheese and bread and were off to fish. This was serious business.

I stayed in our camp of three or four empty stone huts every day to guard things and sketch a little boy who would appear from nowhere and sit patiently for me. We saw very few other Indians and although we spoke no Quechua, they understood oranges, bread, and kerosene and brought us some delicious new potatoes that we boiled to eat with our fried trout. When the anglers trudged back with their fish that first night, it was determined that the only limit, at over 10,000 feet, was what you could clean and then carry back up the mountain from the river. A veritable bonanza, we spread the fish out on the thatched roof at night and by morning they were frozen solid. We packed them in straw, kept them daytime in baskets within the hut and they were still frozen when we left.

Good fishing at Jatunrumi.

We all slept in a circle with our feet to a llama dung fire that someone would keep burning but we were still cold. That first morning when we awakened there were scores of llamas in camp and they watched our every move as we fixed breakfast. The adults were big and powerful looking, but curious and seemingly tame, only batting their long eyelashes occasionally. By the end of three days we had enough trout to fill all our freezers in Cochabamba and were ready for civilization.

On our return we found an enormous boulder in the middle of the narrow mountain road that had a two hundred foot drop on the outside. The winch on the army truck was broken and it took quite a long while for the men to finally nudge the mysterious boulder down the mountain. We saw some geese flying and shot a few but they were as tough as the free-scrounging chickens in Cochabamba. We had flat tires, boiling radiators, fuel pump failures, vapor locks and bad brakes along with the broken winch and this time we could see the precipitous, dangerous road we traveled down with hairpin curves too tight to make in a single turn. But those big and tasty trout were worth all of it.

Ben and Kirby were much admired in the community for their leadership and when he retired, a large farewell party was planned by many of the expatriates as well as the office staff. It was held in the large compound of a private residence and continued for many hours. There was a Bolivian Dance Band as well as a Quechua Indian band that entertained us while we ate with their slow shuffling and strange, sad music. We were allowed to take pictures which we generally wouldn't dare attempt. In our first year in Bolivia, we were sightseeing on the road to Quillacolla with the boys and were chased while trying to photograph some Mardi Gras revelers. This afternoon we were all allowed full use of our cameras.

There was a very old Indian man there, whose son, the band leader, said he was the youngest sibling at sixty and although he didn't know for sure, he just knew his father had to be well over a hundred. His face was a mass of bronzed wrinkles, his eyes Asian, he had a long, hooked nose and with the flaps of his woolen cap covering his ears, it was an absolutely classic Quechua face. Their clothes were all homespun woolens made from the yarn that we always saw the women sheep-herders spinning. Their pants were held up with ropes and their horny feet were in sandals made from old tires.

The *sampoños* were made from reeds, the guitars, called *charangos*, were sometimes made from armadillos, and then there were drums made of skins and other instruments made from gourds. Many of the men were sad to see with their teeth rotted out from chewing coca leaves. The image of that night that will always remain with me was seeing the band players turn to the wall and hunch over their food to eat. There was no communication with each other; they looked as if they were afraid someone would take their food away.

15. Eric: Santa Cruz Office

Our exploration efforts had helped define three areas of interest that met our geological requirements as well as our governmental obligations. We had an agreement with YPFB in the Parapeti Block where we were required to carry out geological and geophysical surveys and to drill a number of holes. Our first test at Mandeapequa evaluated that structure and it was dry. Further obligations to YPFB were satisfied by additional deep tests on structures defined by the seismograph and six strategraphic tests. Although some of these had shows of oil and gas, none of these were commercial. All of this effort had required us to cross the trackless Chaco forest with a network of roads. These were also the roads that Che Guevara used to foment his unsuccessful Cuban style revolution. The area of least interest was the northern region, the Chapare, which was the most inaccessible and not drilled until after I left Bolivia, which was unsuccessful.

The Company's primary interest had always been near Santa Cruz where there had never been any previous drilling. The area was served by mostly dirt roads until a road from Cochabamba was paved by USAID just before we had arrived in 1956. Previously, oxcarts were the only method of transportation during the rainy season. We had seen oil seepage there and had indications of large surface structures that we were eager to test. By this time, in 1959, with our focus on Santa Cruz, we needed to move our base of operations. This was not going to be easy since there were no paved streets, no central water works or sewage facilities even though there were about 30,000 permanent residents and Santa Cruz was the center of commerce for eastern Bolivia. Some of the geophysical and drilling companies already had their offices in Santa Cruz.

The town was like an old frontier outpost, full of suspicious characters after World War II. Both Cochabamba and Santa Cruz had fascinating explorers, men and women, who would travel through Santa Cruz into the *selva*. The town was known for its beautiful, darkhaired women and many European men who had found a haven there. Two Dutch geologists with another company became well known to all as they searched the bars for SS men (the Nazi Military elite), either by demanding to see their arms where an SS tattoo would be, or just ripping up the sleeves of their suspects. Real physical barroom brawls were not uncommon.

My office in Santa Cruz would be in the Gulf compound a few miles from town which was established for all our equipment and personnel. When the office opened we had to request the police to control the crowds that had lined up for jobs. There was tension because the Santa Cruz natives, or Cruceños, who were mostly of Spanish blood, were in competition with the highlanders from Cochabamba, who tended to be mixed with the Quechua and Aymara Indians. Also there was a separatist movement by the Cruceños to create an autonomous state.

We all worried about the problems of schooling and medical attention for the children. We knew that our Juanito had had two successful years at the American school in Cochabamba and almost all the boys' friends were moving with us as well as most of the teachers. We had all found doctors in the Cochabamba community that we trusted, and some dentists. We knew where the clinics were and people that we could depend upon in case of emergency. It would be hard for all of us to give up this security. The company sent a roaming doctor from the States from time to time but that was not enough for life in Santa Cruz. The excitement of finally drilling did not balance our concerns.

16. Libby: Stateside, Astronauts and Rio De Janeiro

Augustina developed a dreadful cough that June winter of 1959 and when I took her to the doctor, he suspected tuberculosis. We were terrified and did exactly as he prescribed. Augustina must go to bed, I must boil all her dishes and feed and take care of her and Inez. We never hesitated a moment, they were family, but after three weeks I returned to the doctor who had never notified us as promised with x-ray results. He was almost jocular and said, "*todo esta bien*" or 'no problem' in today's vernacular, and hadn't we received notification? That plus the amoebic dysentery he had me fighting with small daily amounts of arsenic, brought me to frequent tears.

Eric and I went to see a French doctor who was known for his compassionate nature and sure enough he said, "You smile with your mouth but not your eyes. You are depressed, as many expatriates in Bolivia are", and insisted I go home alone for two months. At first it sounded wonderful but then scary as I thought of all the complications. I finally hired a well known Spanish woman in the community to care for the children and free Augustina to market, cook and clean. This was the second and last time I ever saw Augustina angry, insisting there was no need for this woman, but it made me feel less guilty, and I didn't want anyone to be neglected.

About the only daily news we had was the Hallicrafter radio and our weekly entertainment was Time magazine. We nearly flipped one night when we read that the husband of my best friend Rene, Scott Carpenter, had been picked as one of the seven original astronauts. I had spent several days with them in Lexington Park with the boys before leaving for Bolivia that first time and after this startling news, I decided I would visit them again on this trip. I first flew out west, stayed with my aunt and uncle in Boulder, Colorado, visited friends and was feeling great, especially since I had quit the arsenic! The day before I was to begin my eastern flight back, my uncle wanted to know if I had done everything I had wanted. I told him how we had always dreamed of having a cabin in the mountains, in the same area as theirs, for vacations. He went to the phone, called the realtor in Lyons and we were soon on our way up into the Colorado Rockies.

We looked at four locations but it didn't take me long to see the possibilities of a one room shack with good road access and the sound and sight of the St. Vrain River running through an acre of aspen and firs. My geologist uncle paced it off, checked out drainage, etc., and okayed my decision. I only had fifty dollars in traveler's checks to spare which I offered to Mr. Ramey as a down payment. He gave me a receipt, I signed a piece of paper on the hood of his car and the place was ours. Try that today. Through the years the cabin grew into a house and became our refuge from the hectic pace of life, our little paradise.

After mother in Kansas City and before New York, I stopped for a few days with the Carpenters. What was happening was extremely exciting, not only for the astronauts, but the USA and the world. Scott did his best to explain outer space to me with a rubber ball on the end of a string. One night I was included in a party at the home of the astronauts' manager, consisting only of the seven astronauts, their wives, several German and American scientists and their public relations man, Shorty Powers. Since the Carpenters had to attend a parent teachers meeting first, and since his wife was not in town, I was driven to the party by one John Glenn.

It was a memorable evening that I've often recalled through the years. Powers cornered me to talk about my friend, Rene, the scientists and astronauts were thoroughly absorbing to listen to, and they all expressed much more horror at being in the jungle with big snakes than going into outer space. As these men achieved front page fame through the years I've had much fun speculating which one of them pinched me on my fanny that night as we were all putting on our coats to leave.

Eric and I decided to celebrate our tenth wedding anniversary in Rio de Janeiro on my return trip but that would require a stopover in Lima, Peru. Our former temporary manager and his wife with whom we had shared so much on our arrival in Bolivia lived in Lima. I spent several days with Dave and Barbara in their lovely home with beautiful gardens and house full of kids, and saw all the museums with their Inca and pre-Inca treasures. The striking contrasts were heartbreaking. The poverty of the Indians was similar to our Quechua Indians but the luxurious wealth of Lima far exceeded anything we had seen in Bolivia.

Flying across the South American continent from west to east in a four-engine prop plane, probably a DC 6, over wet green jungle all day long, was positively frightening. However, with only a dozen of us aboard that huge droning plane, I discovered my Spanish flowed more easily after a two months rest. We were all relieved at the sight of Sao Paulo and its millions of blinking lights when we stopped to refuel. Rio was even more warming when Eric met me an hour later.

With one of those freak but marvelous dollar exchanges, we had an extravagant week in plush Copacabana Palace Hotel. We needed separate bathing suits for the luxurious pool and the beautiful sandy beaches. We would return to find them washed and hanging up, our large plush room put in order several times in the day, and in the evening, our night clothes would be laying on a recently made up bed. We never saw anyone; it was almost spooky. We had all the fresh fish we could eat and nightly entertainment with Charles Aznavour in the Hotel or in the many night clubs. Many could speak Spanish but they preferred to struggle with English or our miserable attempts at Portuguese.

Although we had previously made a pact that we would never fly together without the boys, we did on our return to Cochabamba. Due to mechanical difficulties we had to spend the night in a little railroad town called Campo Grande deep in the Brazilian 'wild west'. We slept on the third and top floor of an adobe hotel with no electricity or screens, on a mattress made of grass and with just one sheet. In the morning we were given a cup of very thick coffee, hard bread and rancid butter before departure. We knew we were getting close to Bolivia. All was in wonderful shape at home. Soon Markicitos forgave me for leaving him and Pablito stopped having tantrums when he couldn't have things his way. We were told that Juanito had been a great help.

17. Eric: Drilling, and on to New York

I needed to be in Santa Cruz in the beginning of the drilling program and was lucky to find a spacious and very livable house that was much like an American ranch house. In our first year we had bought a car in Cochabamba but in Santa Cruz, with only roads of sand, it would hardly be usable. We inherited a gardener with the house, Armando, who would pump water daily into a tank on top of the house from the well, pick our papayas, avocados and mangos and keep the encroaching jungle and snakes at bay.

Libby and the boys were driven down to Santa Cruz by Luis who had started with the company on my first field trip to Mandeapequa. Augustina and Inez followed in a truck with all our belongings and arrived at the same time. We soon decided to buy a horse and 'Mando' spent his spare time taking the boys and their friends on and off King.

One night shortly after the family had arrived in Santa Cruz, a geophysical contractor came by and asked if we knew of anybody in the company who had a certain blood type. He had an American employee in the hospital who required a transfusion and they did not have his type of plasma. The employee was returning to the States, had attempted to demonstrate his gun to a prospective buyer and accidentally shot himself in the stomach. I contacted everyone in our company in Santa Cruz but no one had his blood type. The next day I went by the hospital but he had died during the night. I was just plain frightened when I saw the limited facilities and knowing how many people would soon be arriving, I immediately contacted the office in Cochabamba and requested a medical officer for the company.

We had a new manager, Charlie Dresbach, who wanted to run the business from Cochabamba by radio, and he responded by promising to send us the German nurse, Anna Maria, who would be a comfort but not enough for our growing requirements. Larry Perry, a new geologist with the Company, and family, had just arrived. His first job was to map in detail the structural geology in the NW area of our Santa Cruz block.

Next, our toddler, Mark, bit our dachshund, Loompi, on his long soft ear and the dog, in turn, bit Mark on his nose. The septum between Mark's nostrils was split and the only thing Libby could do was apply penicillin ointment, cover the wound with a butterfly bandage and give him a bottle. When I came in that night I found that Libby and Augustina had tied his hands with cotton strips to the side of the bed after he had fallen asleep. They took turns all night sitting by his side to be there when he awakened.

The next morning, a typical gray, muggy summer 6:00 AM, Armando failed to show and so I placed the radio receiver on the backyard steps and then from about twenty yards away began pumping water. I was just getting water into the house when Charlie began demanding "Eric, where are you? Eric! Come in--where are you?" over the radio. I could only shout a few obscenities. Mark's nose healed well but again I demanded more emergency medical assistance.

Several men who were responsible for the drilling operation had just arrived from the States and were to stay at the newly opened staff house. Libby and I were settled now and wanted to have a welcoming party for all these newcomers. I was driving fast in my old, canvas-topped field Jeep at dusk to pick up some of these men at the staff house, when I arrived at an intersection. The car etiquette in Santa Cruz was that whoever got there first had the right of way. I thought I was first but I had cut off another car and forced him to stop. I heard a gunshot and a bullet whizzed just over my head. This was wild west stuff.

I hated to spook these newcomers but I was a bit shaken and wanted to share the experience when I picked them up and besides, they needed to know what life could be like in Santa Cruz. I asked the men not to tell Libby and the rest of the guests and we still managed a good welcoming party that night. Despite the anxieties of all concerned, the move was on and families were beginning to arrive. The hope at this point was that we could find enough oil to make all this effort worthwhile.

The first test we drilled in the Santa Cruz area had shows of gas but tested water. Subsequently we spudded the test of our most promising surface structure that had a gas seep on top. About the same time I was offered a position in Spain and Libby and I didn't have to think long. We had been in Bolivia for almost four years and frankly were beginning to burn out. At the same time, I hated to leave when success might be imminent. I accepted the position in Spain and in a few weeks we were all flying to La Paz with three or four other company men aboard the Aero Commander. The pilot flew us between the Andes Mountains instead of over, which was quite a hair raising experience for Libby and me, but the boys loved it. And although the plane was not pressurized, no one was sick.

Leaving La Paz and on our way to New York, the Panagra pilot flew over a smoking volcano, called El Misti, and said it was intermittently active. I thought of Arque. After Lima, we were scheduled for a short stop in Cuba and I looked forward to some good cigars, but Castro's 1959 coup had just occurred, we were met by armed revolutionaries and threatened with machine guns if we disembarked. We did not try.

Coincidentally, while we were in New York with my parents, I had the opportunity to attend the 1960 Convention of AAPG in Atlantic City, where Libby's uncle from Colorado, Charles Hares, a pioneer geologist in the Rockies oil industry, was to receive an honorary membership. I was invited to a dinner for the honorees and one of these men had worked in Bolivia years before. When he heard that I had just returned from four years of geological exploration there, he commented that we would never find oil in the Santa Cruz area because there was no Tupambi sandstone in that part of Bolivia. He had just retired as chief geologist for Esso and I was intimidated by his dogmatic attitude. I tried to explain that we had high hopes for the wildcat (exploration well) we were currently drilling and there were other reservoirs to test where no one had drilled before.

Two months later in Spain, I received a cable from Paul T expressing great joy. The Caranda structure had tested oil in several different productive sands and would prove to be the first significant discovery in Bolivia since the 1930s. Larry had found indications of a gas seep on the structure. This field produced a sweet crude oil that was connected by pipeline to Cochabamba and La Paz and subsequently to the export terminal on the Pacific coast of Chile. I was thrilled but Libby and I were naturally disappointed not to have been there to share in the big celebrations. Sadly, there was bad news with the good. Our Swiss geologist, Walter, who had been happily married to a lovely Bolivian, had lost his life in an accident in the field.

In 1969 the Company was nationalized after a coup by General Ovando, who was responsible for the capture and killing of Che Guevarra. Gulf Oil never realized a profit from this operation but we all felt personal satisfaction. The subsequent national gas production has also made Bolivia an important source of energy for Brazil and Argentina. Bolivia had been a marvelous experience for me, personally and professionally, and for the whole family. We had learned a new language, the ability to cope in another country, and had made friends from around the world.

18. Libby: Santa Cruz, Farewell Bolivia

The move to Santa Cruz had been in the mill for a long time and it now became a short but interesting, as well as challenging, part of our stay in Bolivia. For us, it was only for three months. I loved our house and beautiful grounds, Armando was a gem, and for a Westerner like me, it was always fun to ride our horse, King.

We were there only in the rainy season and I rode out with him one rare and beautiful sunny morning. Suddenly he shied and I heard the bellowing and constant screaming of cattle. I had come upon a slaughter house in a lovely area of palms and white sand. Since then, even American corn-fed beef has never tasted that good to me.

During our stay in Santa Cruz we seemed to have one health calamity after another and both Eric and I were concerned because promises for medical aid was not forthcoming. I have never been happier than when the offer for Spain was made. The farewell parties and packing up are a bit of a blur. Since no paper was available, the Indians came with gunny sacks, cut them up and sewed them around the furniture. I found a large solid gold tooth filling when we cut away the soggy sacks after our possessions arrived in Spain, and I just knew some poor Indian had lost it while biting into a string.

Everyone had wanted to buy whatever we would sell, but our possessions would be shipped directly to Spain with no opportunity to replace them, and so I was reluctant to sell. Later, when I saw all that could be found in Spain, I wish we'd sold or even given many things away because there was too little of anything in Bolivia for those remaining behind.

Sometimes people have asked me what did we do in Cochabamba? What didn't we do is a better question! We had to make our own entertainment which included puppet shows, theater plays (for which I was the *maquiladora* or make-up artist), costumes, Halloween, and Christmas parties besides the usual dinner parties. We wrote our own weekly newspaper, dug for antiquities, had Easter egg hunts, running competitions (to the winners I once gave some new puppies of ours for which I was not forgiven for a long while), and other parties at the school, hiked and formed a beautiful choir to sing Christmas carols at parties, the school and door to door.

Quechua band, goodbye Bolivia.

At local celebrations, we ate those *deliciosos pero muy picante (*very hot*) saltenas* on the street, while we watched all the colorful dances and the parades with silver utensils sewn onto bright banners on the hoods of cars. They even had a parade for the Union of *Contrabandistas* or smugglers, a thriving business in Bolivia. We also threw water balloons from a big truck at our friends houses during Mardi Gras and other crazy things. We had much fun along with our enormous challenges and responsibilities. I bought a Mexican guitar from a friend and learned a little here and there but didn't really find any satisfaction until I found a classical guitar teacher in Spain. I found almost no time for my drawing or painting, my major in university.

Of course we all struggled constantly to speak that beautiful Spanish language, better. Some had a natural talent for languages, some had great vocabularies and dreadful accents. I am fortunate to have little American accent when I speak Spanish, but I'm always struggling with my vocabulary. And our vocabularies were different. Eric couldn't make his way around the kitchen or market without me but I could hardly speak a word when it came to geology or his life in the field. Finally, the most formidable job of all, parenting without all the stateside aids. My answer is another question: What was this thing called 'television' back home?

It wasn't easy to leave our friends of such memorable times but Spain with all its European culture was beckoning. I had learned more than I ever had in the past or ever would in the future in such a brief period of time. I had become more conscious of my nationality, my heritage and my changing beliefs, and I had also become more aware of the strengths of women, a subject not discussed much in those days. I had met and made friends with many adventurous, competent and sometimes very talented women. I felt a great sense of accomplishment after our long journey in Bolivia. By the time we left, I could express my needs but not my emotions or politics well in Spanish but that would come later in Spain.

I was so grateful to dear Augustina for her endearing friendship and I like to think that in her mind her journey with us was a happy one. We bought her and Inez a little adobe house in Santa Cruz which was little that we could do for all of her devotion for more than three years. She did not want to return to Cochabamba. Although I tried to correspond, the mailing system was miserable, we had been thrown into another new busy kind of life, and Augustina couldn't write. She worked for some friends for a while but oil production took over, people moved on and we lost touch.

We heard Inez had been educated in a business school and grew into a very capable young woman. I have never wanted to return. Our experiences and all the accompanying emotions are best just remembered. Due to cocaine, oil and agriculture, Santa Cruz is now a metropolis of over a million people and they say we wouldn't even recognize Cochabamba. However, the Morales Clinic where our Mark was born is now THE Clinic in Cochabamba.

Part II
SPAIN
1960–1963

19. Eric: Colorado Rockies, and Hello Spain!

The family and I had two months of vacation time for every two years in Bolivia. Those two months were needed to include wrapping up my business in Bolivia in the Gulf's New York office and finding out what I could about the geology of Europe and Spain in particular. Gulf's New York office at that time was responsible for Latin American exploration operations and had no data on Europe. I was challenged by the fact that there had been no production of oil or gas in Spain. However, I had excellent contacts with the geological staff in New York, partly because I had worked summertimes in the office while I was studying geology in Colorado.

My stay in the States not only included the AAPG Convention in Atlantic City but also the much anticipated trip to the Colorado cabin that Libby had bought the previous year. The boys and I were thrilled and delighted with the cabin, the surrounding mountains, forests, and the cold river with trout. It gave me a great feeling to have a place of our own where we could vacation and build upon in the future. The boys wanted to climb the mountains immediately and although we had to set up some rules, we did not have to be concerned about tropical diseases, bad water and revolutions, just a few ticks.

While in New York, I received a letter from Steve Davies, the newly appointed American manager of the operation in Spain, called CIEPSA. The letter explained some of the operation that was underway as well as information about the life style in Spain and suggestions for what the family would need to bring. The organization would be a combination of three companies: the Spanish Company, CIEPSA (*Compania Investigadora Petrolifero de Espana*), the German company, Diehlman (a company with coal and oil production in Germany), and

Gulf. CIEPSA and Dielhlman had been working as partners for 10 years and had drilled a number of exploratory tests that had shows of gas but no production. The Spanish had the exploration permits from the Government, the Germans brought their own drilling rigs, and Gulf was bringing money. We were all contributing personnel to the group effort and the American manager would be in charge of the whole operation.

We would attempt to carry out an efficient technical operation of exploration and drilling as well as the financial and executive functions in three languages. We would also be operating in the Basque region of northern Spain with its own language and culture. It passed through my mind that besides the fact that I spoke Spanish and was a geologist, the company may have taken into account that I had grown up in New York City and had survived the streets of Brooklyn. My neighbors and friends had come from households of mostly European languages with their various temperaments. I had confidence in my survival skills in Spain and was looking forward to being part of this European stew of personalities.

We decided that Libby and the boys would stay behind in the States while I went to Spain, find a place to live and begin my work. I left In June on a TWA night flight and awoke the next morning to see that we were just crossing the coast of Portugal on a bright sunny day with no noticeable pollution. After customs, a company representative met me at the airport in Madrid, checked me into the Wellington Hotel and then walked me to the Spanish Gulf office. This office was responsible for operations in mainland Spain as well as the Spanish Sahara (now Mauritania) and the Canary Islands. I was to be the Geological Supervisor to oversee and administer the geological operations for CIEPSA in the northern Basque region.

I met the American manager of the Madrid office who briefed me on the operation in Vitoria. He showed me maps of the different exploration areas which were mostly in the northern Basque country in the Cantabrian Mountains. He was happy that my family and I spoke Spanish and began to brief me on some of the people I would be working with in Vitoria. He stressed how important it was to be able to communicate technically with the Spaniards and to get along with the German personnel. There were currently three drilling rigs operating and one of them was testing an apparent gas discovery. After meeting administrative people in the office, I was given a ticket for the Talgo Express train that left the next morning for Vitoria.

Basque country in the Cantabrian Mountains.

On arrival, I taxied to the Canciller Ayala Hotel and was picked up the next morning by Steve on the way to our office. The office was situated in an apartment building that included private residences and I was introduced to most of the personnel, including the exploration manager, who was my immediate boss, Ossie Schmidt.

He was of German origin who had worked in the Ploesti oil fields in Romania during World War II and began to work for Deihlman after the war. We would have a layer cake of American, Spanish and German personnel too involved to try to explain. The geological and geophysical work would fall under Ossie. I met the two other American Gulf personnel in the operation, Art Van Damme, the geophysicist and Garland Findley, the drilling engineer. Neither of their wives had arrived yet, one Italian and one British, and I decided to rent an apartment in this building until Libby and I could find a permanent home.

The Canciller Ayala Hotel was new and overlooked the City Park which had a round band stand that played an orchestral recital every Sunday. While staying in the Hotel, I was sitting at the bar one evening, enjoying a glass of *tinto* (red wine) and reading the international Herald Tribune, which was published in Paris. It was truly a global paper in its coverage and one of the few English language publications allowed in Spain by the Franco Government. An English voice interrupted my reading and apologetically asked if he could borrow part of the paper. He looked familiar to me and he explained that he was a Shakespearean actor who had begun in movies playing the bad-guy in Tarzan pictures. It was Anthony Quale and he had recently completed *The Guns of Navarone* and was en route to Madrid to work in the movie, *The Decline and Fall of the Roman Empire.* It was obvious that Vitoria was going to be more of a cosmopolitan city than one would have expected of a provincial capitol in the Basque country of Spain.

There would be several German geologists and two Spanish geologists who would be working for me. The Germans had some operational experience and spoke fairly good English. The two Spanish geologists had no prior experience in petroleum exploration and their English was limited but all of them seemed agreeable and friendly and I was enthusiastic with this new challenge. I was immediately thrown into an environment of three companies, three languages, three drilling wells and geological challenges in the beautiful capital of Alava, one of the four Basque provinces. The fourth language, Basque, was a mystery to all of us. I had a month to get an overview of the operation before returning to the States to get Libby and the boys.

20. Libby: Madrid and Vitoria

Our flight out of New York to Madrid in 1960 was pure luxury. The Company decided that it would be less expensive to fly transferring families first class with unlimited baggage weight than to pay the extra baggage costs that accrued when we were flying tourist class. But we would still fly tourist on vacations. And of course our dachshund, Loompi, had to be checked through in a cage. We hand carried valuable possessions, including silver flatware and photographs that we didn't trust to the long ocean voyage, enough clothes to cover changing climates, favorite stuffed animals, my guitar, cameras, Eric's business necessities, etc. Our household affects would travel by truck from the jungles of Santa Cruz, Bolivia, to Cochabamba, then by rail over the Andes Mountains to Arica, Chile on the Pacific Ocean, and then by boat to the Panama Canal and into the Atlantic Ocean and eventually to Spain. We hoped! We had just spent two months with our kind families and friends in their extra bedrooms and roughing it in our Colorado mountain cabin without electricity and water. Now, in our first class flight from New York to Spain in August 1960, we were experiencing what I considered to be a well deserved reward for our challenging years in Bolivia.

At Idlewild, we awaited our flight in the first class Club Room, with attendants meeting all our needs. After we boarded, our busy boys were entertained, fed and practically danced down the aisles while Eric and I sipped champagne. The stewardesses played card games with the boys and escorted them to visit the pilots who spun stories and gave them airline pins for their lapels. Quite a difference from my flying experience from New York to Bolivia. We were served gourmet food in several courses that would have satisfied most anyone eating in a first-class New York restaurant. Finally our chairs were laid back to almost lying, we were surrounded with blankets, down pillows, eye covers and with only the drone of the engines, we joined our boys in slumber-land.

In a photo we still have, Spanish photographers captured us descending the plane in Madrid and after noticing our South American accents, they gathered around asking questions. The boys were wearing their navy blue jackets, white shirts and bow ties, very different from today's uniform of jeans and T-shirts. The only thing that distinguished them as Americans were their burr haircuts, the trend in Europe being longer hair and bangs for boys. The photographers shoved pictures into our hands, protesting any compensation and Eric and I were quite surprised with all the attention.

Custom officials were quite formal and polite while thoroughly checking through all our baggage. You can tell a great deal about a country by how the custom officials treat you. I immediately felt surrounded by Spanish culture with all its dignity and grandiose history. They picked through my lingerie, as this was a Catholic country and contraceptive aids were forbidden, but nevertheless available under the counter. Spain, like Italy, had an unexplained low birthrate. When I questioned an Italian friend some years later about this phenomenon in her country, 'The Catholic Country', she said it was simple, "We just looked at all the people, the growing pollution, how difficult and costly life was becoming, and asked ourselves, is the Pope going to raise all of these children?"

The Guardia Civil (the government police), all appeared to be handsome, tall men, with classic Spanish features, wearing tricornered black hats and watching every newcomer very closely. We knew we had arrived in a world totally controlled by the dictator, Francisco Franco, called *El Caldillo*. However, this also meant it would be a safe country for us to live in. It was also comforting to know that all of us Ericsons could 'defend' ourselves in Spanish, as the Spaniards would say, and the boys immediately felt at home, probably as much as in the USA.

Their 'at home' attitude in the Spanish language with South American accents plus blonde hair and outgoing mannerisms acted as a magnate and they helped us sail through the next three years. They loved Spain as we did and no matter where we went, Loompi was accepted with much affection.

After a long nap at the residential Wellington Hotel, we walked over to the luxurious Castillano Hotel for high tea, as it was the only food thereabouts at that time of day. Sitting in the huge lobby was the glamorous and beautiful Ava Gardner in a high-backed carved chair with her mother close by. The boys were only interested in food but Eric and I were mesmerized with Ava's creamy white skin and classic features, even if she was 'smashed' as the concierge intimated. During that first year on our visits to Madrid, Eric and I would drop by the Hotel Suecia in fruitless attempts to sight Hemingway. These were soon to be the days of Jack and Jackie as well as the Beatles. Movies like *Hatari* and *The Pink Panther* were being made and both skirts and hairdos were high. On one trip to Madrid we found the plains covered with tents and flags, horses and men all dressed in 11th century clothing and armor for the filming of *El Cid* with Charlton Heston and Sophia Loren.

Our solution to jet-lag those first few nights were walks along the Paseo del Prado that entered into the Plaza de Colon (Columbus). At midnight it was still alive with young and old, as well as families pushing prams with wide-awake children, laughing and chatting with friends. We had had some introduction into late nights with local families in South America, but nothing prepared us for this. Children accompanied their parents into bars and while the adults had a glass of *tinto* at about 10 cents a glass and ate *tapas*, the children might have lemonade or a sip of *sangria* with their *bocadillos* (sandwiches). The bars were like a family gathering place and the bartenders and patrons were like the friends and neighbors we would know in this country. On that first visit to Madrid, we often ate in elegant old restaurants with such realistic still-life paintings of bread, cheese, eggs, fruit, etc., decorating the white stucco walls, that I was soon able to drag Eric and the boys with me to visit the poorly lighted El Prado Museum. It was absolutely jammed with the paintings of Valesquez, Goya and El Greco, to mention a few.

After several days in Madrid, we boarded the Talgo train for Vitoria where we were to live for three years. It is a beautifully picturesque town that can rival almost any town in Europe for its antiquity. We had a two-week stay in the local Canciller Ayala Hotel, an important fixture in the community. The boys loved to order oranges and watch the head waiter, who, with an exaggerated flourish for them, peeled and sliced the orange without once touching it with his fingers. We were frequent visitors through the years and on our last trip in 1986, twenty three years later, they had still not forgotten us or our sons names and the Talgo from Madrid was still a delightful journey. Situated on one of the historic routes between Madrid and Paris, Vitoria was the overnight stop for travelers from all over Europe. We often saw Hollywood movie stars there, trying to escape their fame in the USA and the big cities of Europe.

Soon we were temporarily housed in a spacious, completely furnished apartment with large windows overlooking this lovely old Basque town of 80,000 with cobblestone streets and ancient buildings. The Basque family that owned the building gave us a very warm welcome and even the miserable elevator that was constantly being repaired had a uniformed, proper and friendly operator who always managed to joke with the boys and help me with my packages upstairs, despite all his tinkering.

One night Eric had to leave for the 'well' immediately after supper. The boys were in bed and I was washing dishes when a very thin wine glass broke in my hand and cut me badly. I tried to stem the blood but it was obvious I had some very deep cuts. I wrapped a towel around my hand and walked downstairs to the French couple who were with Schlumberger, a service company for CIEPSA, and renting an apartment below us. When Henri Moskowski saw my hand, he sucked in his breath, muttered in French to his wife, Danielle, and drove me to the hospital immediately. He told Eric later that he was reminded of WWII when they sewed up wounds with no anesthetic. There were four cuts and the worst took nine stitches. The young Spanish doctor used calming words but they seemed to be as much for him as for me. He used what looked like a large sewing needle and it took all my self control not to appear like a self-indulged, helpless American woman as Hollywood so often depicted us, or cool and calculating, always morally loose. These were images that those of us living foreign had to confront. As for this experience, it was an introduction into the medical facilities in Spain; not too super, but far different from Bolivia.

Right on the heels of this episode, John, who was not only well coordinated but fearless, decided he would climb down from our seventh-floor balcony and pay a visit to his father in his office on the second floor. One of the caretakers spotted him, set up an alarm, and soon the office employees, along with us terrified parents, were hanging out of the windows urging this wild American seven-year old to climb into the nearest balcony. He had just discovered mountain climbing at our cabin and heights only seemed to challenge him. The Spaniards thought he was something else.

We were soon invited to have dinner with Jose and Segunda Areitio, owners of the building. These were Basque people, brusque but warm and friendly, and two of their eldest daughters were secretaries for our company. There were seven children in all, and the youngest, a beautiful little girl with dark hair and pale skin, about three like our Mark, generally had to eat in the kitchen with the maids. However, an exception was made and she was allowed at the table with Mark. They were both exemplary. We were asked about lifestyle, politics, and philosophy and when the teenagers were given permission to speak we heard an unflattering and animated conversation about American politicians. The situation between the Basques and Franco was only alluded to, but we discovered that one of Jose's brothers did not speak to the other two, due to their political differences. We also learned that

Franco was mentioned only carefully and since Hitler had supported Franco during the Spanish Civil War (1936-1939), Nazi sympathizers were not uncommon in Spain.

After my hand injury, maids were immediately sent for interviews and I hired Isabel. She was a small attractive woman in her thirties, the mother of two young teenage daughters, always bustling around. She became a fixture while we were in the apartment and stayed with us when we moved into our rented home. It took another four months after our arrival for our furniture and possessions to arrive from Bolivia, having mistakenly been sent to Barcelona on the Mediterranean Sea instead of the port of Bilbao on the Bay of Biscay! Our shipment had to sail around Spain, through the Gibraltar Straits, past Portugal and around the western part of Spain to eventually find us. It was heartbreaking to see the damage when the large crates were opened. The furniture was wet and the beautiful hand-woven Alpaca carpets had been drenched with sea water and were falling apart. I lost heart after losing some of the ethnic treasures we had acquired and vowed never to collect again. However, we had some pieces made of South-American walnut that we had refinished and still treasure.

21. Eric: Spanish and German Work-mates

That first month in the office had been a blur. Most of the office was Spanish speaking, English was understood by the majority, and the Germans spoke English better than they spoke Spanish. It took quite a while to begin to understand the personalities and my position in the organization. This blend of three cultures was completely different from all that I had experienced in the States or Bolivia. The Spaniards were reserved, the Germans were cautious, but both were respectful to all.

At this time, the Spaniards were recovering from the Spanish Civil War and the Germans were dealing with their defeat in World War II. Franco's regime had been supported by Germany during the Civil War against the Spanish Republican Government from 1936 to 1939. The Government had had support from the Soviet Union. The Basques had supported the Spanish Government against Franco for promises of more autonomy for the Basque region and after the war Franco punished the Basque by forbidding them to use their language in public affairs, or to teach it in schools. Also, part of this mix had an undertone of the Spanish American War in which Spain had lost Cuba, Puerto Rico and the Philippines. This ended the old Spanish empire in 1898 and was the backdrop for the atmosphere through which I was trying to find my way. I soon became aware that most of the office was supportive, at least outwardly, of the right-wing Franco government. Criticism of the government was unspoken. No one talked politics.

In the geological department I was supervising four geologists: two Spaniards and two Germans. We were also supported by a Dutch paleontologist. The Germans had some experience in drilling operations but the Spaniards had none and had just recently graduated with degrees in mining geology. American oil geologists concentrated on the kinds of rocks, or lithology, and it seemed to me that the Europeans concentrated more on the age of the rocks, or paleontology. The trick was to integrate the two in the most productive way. I also had a feeling that the Europeans were over-dependent on 'the book'. I had had enough experience at this time to use 'the book' with caution and be prepared to innovate.

On my return with the family, I was immediately occupied with the evaluation and completion of what appeared to be a natural gas discovery. Preliminary tests showed one of the wildcats could make a commercial gas producer and this would be a first and very important discovery for Spain. The problem was that the gas was found in 3000 feet of fractured rock and it could not be definitively tested before casing the hole with pipe. The final decision took months of discussion and in the meantime, we were constantly concerned with keeping the gas flow under control with heavy mud since it was trying to blow out. We finally decided to run a perforated liner through the whole section which was successful.

This well, Castillo I, was situated about fifteen minutes southeast of Vitoria on the edge of the Cantabrian Mountains. I made this trip day and night with several of the geologists, the exploration manager, Ossie, and the general manager, Steve. It was an exciting time and I was able to become more acquainted with almost everyone involved. I got along well with Ossie and began to develop a close relationship with one of the Spanish geologists.

It was obvious that the owner of the apartment building welcomed the American group with CIEPSA that had offices in his building. He also benefited from the apartments that he rented to foreigners that were involved in the oil business. Shortly after we settled into the apartment, I was invited by Jose to join him, two of his friends and Henri, to go to a Jai-Lai match, or *Pelota*, between two star players. Jai-Lai was originated by the Basque, similar to our handball, but it is played with a basket that fits on one hand as a glove. This apparatus accelerates the speed of a hard leather ball and can be dangerous. It is played off all the walls in a court, also known as a *fronton*, as well as the ceiling, and is physically demanding, requiring both strength and agility.

Castillo I gas test.

First we drove north to Beriz for supper which was memorable due to the quality and quantity of the steak. It was bought by Kilo weight; small, medium and large. It was smothered in garlic when grilled over a hot fire to your choice of doneness. This was a small restaurant and Jose was obviously a frequent customer and a well recognized former Jai-Lai star. The meal was excellent, served with *tinto* and a salad, flan for dessert, followed by coffee and a cigar. After supper we drove to Durango, further west, and Jose told us a little of his experiences as a professional Jai-Lai player in the 1940s and '50's. He and his brother played in Cuba as well as in Miami. He mentioned that Hemingway enjoyed Jai-Lai and was often in the audience in Cuba. When we arrived at the match, we found the place jammed. Betting continued throughout the match but it was all too fast for me to comprehend.

On the drive home, Henri talked of his experiences with the French resistance in World War II and most of our conversations were carried on in English. The whole evening was memorable and a great pleasure for me. It was late when I returned home to our apartment and the family was asleep. I carefully made my way into the bedroom and then bed, thinking I had succeeded in not waking Libby. However, before I could drop off, Libby was awakened with the smell of the garlic. We had window coverings called *percianas*, a great invention of wooden slats, like a venetian blind, only outside the window and operated by chains. Great to keep out sun and noise but also fresh air. Even after we opened the *percianas*, Libby claimed she was still suffocating from the smell of garlic.

Both Libby and I were finding that our Spanish was improving dramatically since there were very few Americans in Vitoria with whom to associate. The office opened at 8:00 in the morning, was closed from 12:00 to 2:00 for *la comida*, which was dinner, then closed at 6:00 PM. After the office closed, people would often go home, change clothes and walk around town for their *paseo*, which was essentially a social hour (or two). On the crowded main street, called El Dato, the men would play cards on sidewalk tables morning, noon and night. Weather permitting, that is. Always carrying umbrellas, Libby and I would occasionally go into the bars for a glass of *tinto* and *tapas*. One bar had sawdust on the floor and they grilled fresh shrimp and sardines to your taste. All the bars that lined the street were noisy, smokey, smelled of garlic and olive oil, had their *tinto* in wooden kegs and *jamon Serrano* (smoked ham) and *chorizo* (hard sausage) hung overhead.

New friends and acquaintances would entice us into the bars

to enjoy our South American accents and discuss the subject of the work going on at the well, which was of interest to everyone in the community. Although the people socialized in bars instead of at home, and excellent red wine was cheap, we never saw a drunk or a bar fight. It was a comfortable place to live, not too big a town and the people were always responsive, friendly, and helpful. We immediately felt *en casa* (at home).

Now Libby and I were faced with the necessity to buy a family automobile. I had been assigned a Peugeot 403, a French standard, postwar, middle-class car for my company work. We discovered when we began to look for a personal car that the Peugeot 403 was reliable, comfortable and relatively inexpensive. The other consideration was that in selecting a car, I had to remember my status on the organizational chart. This was most un-American but very important to the Europeans in the company. Steve's company car was a sleek, black, new Mercedes 220. Klaus drove a 190 Mercedes, and Ossie drove a 180 Mercedes, the lowest price model. It would have been socially unacceptable for us to buy a car equal to or better than Ossie's. The German geologists drove Volkswagens and I had to find something in between. So we bought the Peugeot.

22. Libby: The Neighborhood and Bullfights

Our house was in a section of Vitoria called Ciudad Jardin (Garden City) of maybe thirty houses that were considered summertime challets. During the winter, most owners lived in apartments downtown where the heating was not so costly. Economically, times were not that good in Spain and many rented their homes to foreigners, although we were few. The exchange rate was about sixty pesetas to the dollar. There were almost no imports and I doubt many exports. Franco wanted his Spain pure and independent of the rest of the world. We paid fifty dollars a month for this enormous house with a third floor that we never used and we still lacked enough furniture to make it seem cozy. This was a house built for summers and many servants.

A raincoat and umbrella were absolute necessities in Vitoria. We had cold, damp weather for most of the year and light snow in the winter months. I soon hired Ricardo, a tall, handsome and gentle Spaniard, the father of three daughters, who had fled southeastern Spain during the Civil War. He would come early every morning to fire up the furnace for the radiators. He also took care of our large yard and garden which grew our fragrant carnations, considered a 'must' to throw to the bullfighters, mushrooms and snails for favorite omelets that my landlady, Blanca, taught me to prepare, and a huge cherry tree, all behind a wrought iron fence and a thick tangled hedge. It wasn't long before Ricardo was raising rabbits in a hutch to sell for local consumption. The boys loved to cuddle those soft little creatures and Loompi was positively amorous about them. I don't know how Ricardo handled their periodic disappearance with the boys but they probably had some idea and were too fond of Ricardo to complain.

Bill Matheson was brought by our company from Scotland to

Vitoria to teach our John and the two sons of our American manager. In his late twenties, Bill was tall, fun, well educated, with a delightful Scottish accent and youthful mannerisms. Our boys thought he was the greatest. All three of his students had been studying in the southern hemisphere and he brought them up to where they would be in the northern hemisphere. I put Paul into a French kindergarten down the street but incredibly, the Nuns tried to get their charges to copy one letter of the alphabet continuously on a 3 by 5 card the first morning. That was the end for Paul. After several attempts to catch him as he ran around the house the next morning I broke down laughing. I reasoned there were too many opportunities to educate him in Spain and soon had him reading enough for the first grade.

The boys immediately made friends in the neighborhood but especially with the children next door. They had a big German Shepherd who could devour Loompi in one gulp, doting parents, cousins and aunts, plus a television! The Spanish women, true *dueñas de la casa*, sat around a round table with an overhanging cloth and a copper brazier underneath for warmth, which the servants kept filled with hot coals. After school these women sewed, knitted, gossiped and drank tea while supervising the young ones as they watched TV. It was the first regular TV our sons had seen and late afternoon TV catered to young people; it was a safe world into which our boys could escape daily. Only once did we experience any embarrassment with these friendly and generous people. The Señor of the house asked Eric if he would stop our sons from shooting off cap pistols; cowboys and Indians being the entertainment for little American boys at the time. It seemed the noise brought back too many vivid memories of their recent Civil War.

Our house was on a corner and directly across the street was an Order of Seculares. The women wore ordinary clothes but lived as Catholic nuns. They were in service around the world and obligated to serve in the community wherever needed. They would have adopted our towheaded three year old, Mark, if we had let them. He rode their big horse while they plowed their land and shared afternoon tea with him and yes, they spoiled him dreadfully. The local Basque men stood on the corner at our gate every afternoon smoking pipes under their black berets and muttering in their guttural Basque language. When Eric returned home one evening, they told him to beware, or the nuns would turn Mark into a *padre*.

The Ciudad Jardin was on the edge of town and just across from us, next to the Convent, were some low, rounded hills with no vegetation and one flat one called the *Monte de Tortilla*. These hills were covered with bicycle and motorcycle tracks and after school and all day on weekends, it was forever busy with little to big boys racing and jumping over the hills with their bicycles. On weekends, young men on motorcycles came to compete and all the little boys watched in awe. Our boys were no exception, each getting new *bicis* for their birthdays. We became friends with the Beistegui family, the local bicycle manufacturers, whose products were sold throughout Europe. With their friends, John eventually became known as the soccer player while Paul became the bicyclist. Before we left three years later I think Paul had mastered every hill. We'll never understand why he didn't break a bone.

Three New York photographers, who were old school chums, met in Paris from different European jobs and decided to travel together for a holiday in Spain. Driving through Vitoria early one morning, the driver, like his friends, fell asleep and hit a telephone pole. We were called to the hospital to translate for a young woman with a concussion and a young man with a broken arm. They were most grateful and later we were invited to one of their apartments for dinner in NYC, but eventually lost touch. As a result of this hospital episode, I was contacted by Berlitz Language School who needed an English teacher. However, when I opened my mouth at the interview, I spoke with an American accent, not British. They most apologetically decided I would not do.

Late that summer was Vitoria's celebration of the Virgin Blanca, their patron saint, and we were expected to attend the first bullfight. Our party included our manager and wife, stateside visitors, and several Spaniards. When we saw our first kill, a couple of women left, causing smiles, but I hung in there trying to determine what it was all about, knowing that I was being watched. Theoretically, when the *picadores* would stick the bull in the muscle of the neck with their lances, supposedly to be almost painless, he would drop his head, allowing a clean kill over the front of his head and into the neck by the matador and the bull would drop immediately.

Ordoñez and the bull.

It is an exciting spectacle to watch with the flashing red capes and 'suits of lights' ballet dancing around the arena. I can never forget the shouts of OLE! and the live music with its *pasadoubles* and trumpets announcing each *cambio de tercio*. During our three year stay we watched some of the most famed of the bullfighters: Dominguin,

Ordonez, Paco Camino, El Cordobes. Most kills we remember were quick and clean and afterwards red flags waved in town where fresh bull-meat was sold. I would prefer the Portuguese method of fighting the bull but not killing it. But having heard the bellows of cattle being slaughtered in Bolivia and reading what happens in waiting pens and slaughter houses today, it is all a debatable subject.

During Easter break we drove down a blossoming Mediterranean coast with the boys and Bill, John's teacher. At one place we watched *paella* being prepared on the beach in an enormous *paella* dish with fresh seafood. The women were thrilled with our interest and insisted we share. I've never been able to duplicate that *paella* which I blame on the lack of sea air and sand. In Alicante, we stayed in a hostel and watched the Easter religious celebrations. One memorable night we watched an exceedingly quiet and erie procession of masked men in white robes with high conical hats (reminding us of our Klu Klux Klan), while others carried a statue of the Virgin Mary, the cross and various relics, all by candlelight. Further down the coast, it was startling to see a Roman amphitheater in Segunto and then to read history of how, previously, the people of the town jumped off the cliff to their death rather than surrender to Hannibal. Bill would read the history as we drove along and a curiosity for other worlds and interest in other cultures was born in our sons.

Weekends were often an exploratory adventure with the ritual of large meals and always with the boys, Loompi, and often a visiting bachelor. The countryside was dotted with uninhabited, crumbling castles often under heavy gray clouds on some inaccessible hill and surrounded by carefully tended vineyards. Every village in northern Spain had Inns or Cafes with tables sanded down each day and topped with hearty food. Our Spanish friends pulled out their maps and guided us through the heavily forested and mountainous country to the best spots in those old villages. At most places, after a light soup, there would be a leg of lamb or chicken pulled out of a brick oven and served by the owners, with the favorite fresh or canned cold asparagus topped with cream dressing, then a salad, like the French. Baked livers with eggs was another specialty. Their canned fruit or flan followed and men in this very male world were always served brandy and a cigar, generally compliments of the house, despite this being a poor country at the time. *Tinto* was served us and the boys drank water or lemonade.

We all loved these trips and I had been exposed to enough guitar music in Bolivia to know I wanted to pursue classical guitar lessons. I found a guitarist who also taught and convinced him I was serious. However, he decided I should learn a popular method called *por sifra*, with each line indicating a guitar string and marked with notes, so besides cords and picking up songs on my own, I only learned to play the classical and flamenco that he wrote for me. I always regretted that I never learned to read the true music form, but given my busy life then, it was probably all that I could handle. In any case, for the next twenty years I was able to entertain myself, my family and friends, and even a few small audiences.

23. Eric: Castillo I and Cueva de Altamira

I was becoming familiar with the other Gulf expatriate staff besides the manager. Art, the geophysicist, a Belgian naturalized American citizen married to an Italian, Lisa, also reported to Ossie. Garland, the petroleum engineer, an American married to Audrey, from Britain, had transferred from Mene Grande, the Gulf Company in Venezuela. His supervisor was Klaus, a German, with Diehlman, and head of the Drilling and Production Department. I was told that he was a Panzer Captain in the World War II, and the Germans said that he drove his Mercedes like a madman to San Sebastian every weekend to be with his wife. Ossie and Klaus were on the same organizational level, with Art and me reporting to Ossie and Garland reporting to Klaus. Steve, the American general manager of the three companies, shared the executive responsibilities with Rafael Teijero. Rafael was from Santander and was the assistant general manager of the CIEPSA group.

In addition to this cacophony of languages and cultural confusion, we had an ever-changing number of consultants on temporary assignments from the United States, Germany and Britain as well as geophysical crews. This sounds like an enormous operation and in some ways, it was. There was a lot of work going on with one or two rigs usually drilling in the field. However, after ten years of dry holes with interesting shows of natural gas, the Castillo well would be the first that might have commercial possibilities. As such, it was extremely important for Spain, which greatly needed the energy supply to supplement its coal production. Spain was envious of the oil and gas production in the Aquitaine Basin in southern France. Natural gas in Europe was a treasured energy commodity, unlike the United

States where most oil companies regarded it as a curse. There was abundant natural gas in the States but with few long-distance pipe lines, the gas just had to be flared. At that time one thousand MCF (a unit of sale) was worth 5 to 10 cents in the States and in Spain it was worth 70 cents to a dollar.

One of the first trips we made into the countryside with the boys was the famous site of the Cueva (cave) de Altamira west of Santander. My interest in history and archeology accompanied my interest in geology. I was always on the lookout for artifacts and I knew that the cave at Altamira dated back to middle stone-age. In Spain there was a continuum back to the earliest of the human species and predecessors like the Neanderthals.

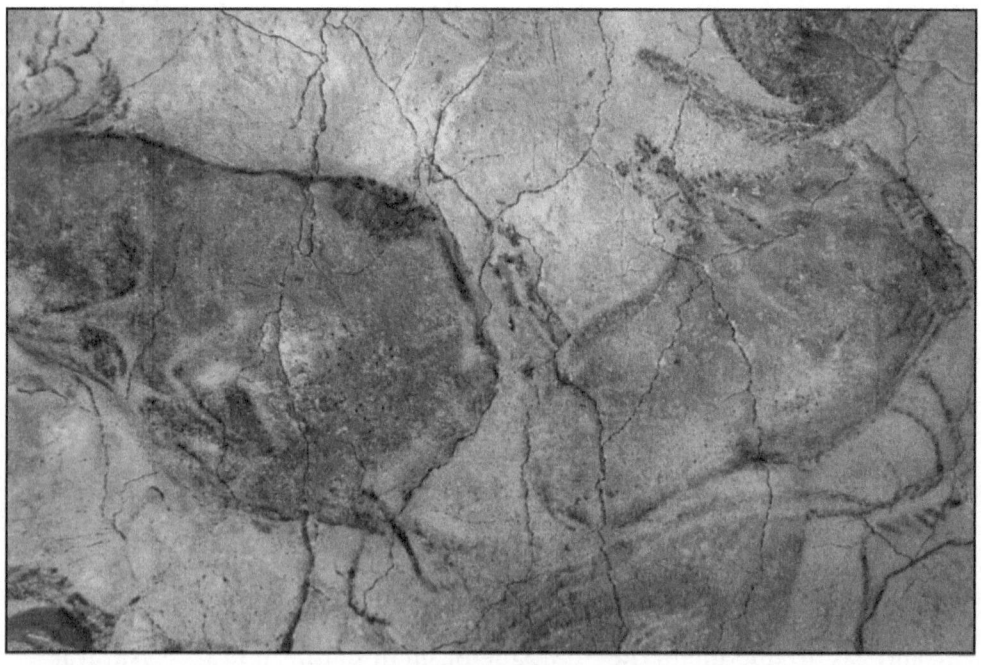

Cueva de Altamira.

9th Century well in Santiana del Mar still works.

This was in relatively open countryside, just south of Santillana del Mar, a village that the Government designated to be maintained as it was in the early 1500s. It was a charming village, dating back to the ninth century, with an ancient water well and medieval church in the center of town. There were no other visitors when we arrived at the cave but a caretaker appeared and agreed to guide us. We followed him and what I remember most was the incredible color of the paintings that had remained after millenniums and the rocky protuberances that depicted a bison's shoulder and/or rump. We were not able to see any of this fantastic art work until the few electric light bulbs that lighted the passageway were turned off, and then an oil-lamp was the only light in the cave. We had to lay on our backs in one alcove to see the drawings. We could also see the black smudges from burning animal fat that had allowed these ancient peoples to see and paint. Now this cave is closed to the public but there is a replica of one alcove that contains copies of these paintings.

The rock formations in the Cantabrian region west of the Pyrenees contained many other caves that also had evidence of stone-age occupation. This particular cavern that we visited was discovered in 1879 and when the paintings were found and the news published, French archeologists thought the Spanish had created something bogus. During World War II, French schoolboys found a cavern near Lascaux, north of the Pyrenees, that contained many more paintings. Since that time there have been other discoveries that may be as much as thirty thousand years old. In the mid-eighties, Libby and I visited a replica of the painted chambers at Lascaux, France. Impressive as these replicas were, they could not match the experience of being in an actual site where our ancestors had expressed themselves so beautifully long ago.

Soon Steve told me that Max Littlefield, who was Gulf's roving stratigrapher, would be arriving shortly for the summer. The company was concerned that we really didn't know enough about the lithofacies, the different kinds of rocks in the area. Max was considered to be one of the top analytical stratigraphers in the United States and all the available well cuttings here were at his disposal. He would give us a new insight into the geology and was much respected by Ossie. Max did not suffer fools easily. He had been a lightweight boxer in college and had a short temper. Libby recalls his anger at the elevator operator in the Canciller Ayala Hotel who didn't understand his directional order of *dos* for the second floor instead of *segundo*. He had worked in Cuba and figured he knew enough Spanish. We became friends and he was a frequent guest in the house. It was a friendship that lasted long after he retired.

Max, along with Gerhard Bivank, the paleontologist, helped me to develop a mental image of the entire stratigraphic column, the rock section from top to bottom in the Cantabrian Basin. We were exploring two and one half million acres; much of the surface geology had been completed and the aerial photography was being studied. The structure was complicated on the surface because it was on the western plunge of the Pyrenees Mountains and further complicated by salt domes. Despite favorable sections of rocks comparable to productive sections in France, nothing commercial had been found in Spain. Max reasoned that there were possibilities, but along with the rest of us, could not figure out why the Spanish section could not produce more gas. By the end of the summer, Max had worked out a facies concept that seemed reasonable and detailed enough to give us new prospect ideas.

We completed the Castillo I gas well and made locations for subsequent tests in that vicinity. We also had to make exploratory tests in areas for oil or gas using the new ideas generated that summer. This meant that I had to travel with the geologists in the mountains and basin areas in northern Spain, roughly two hundred miles from the eastern border with France to the west of Bilbao, and a hundred miles south to the Ebro Basin. Although the weather was often rainy, the geology was fascinating, the company good, and trips were organized not only around where drilling was occurring, but as well as where seasonal food and the cuisine was best.

Slowly I became acquainted with the well-site personnel. The Germans had developed a system of training geological assistants from the local communities for doing work that did not require a college degree. The drilling was slow and a sample was taken every meter from the mud returns and their job was to examine the well cuttings. The Germans and Spanish both required a great deal of detail and expected the work to be done meticulously. After about six months, I was convinced that they had gotten every bit of information possible to discern from the cuttings. Along with the data that Max had provided I decided that there was little more they could do. My challenge was to put it all together from a fresh perspective.

24. Libby: Our Cabin and Spanish life

We were now scheduled to return to the States every year for a one month vacation and this made our foreign treks considerably less traumatic. Our families were very happy that we were in a European country now. My professor sister made sure I had connected with the Calvert teaching system to enable me to oversee the boys education and to ascertain that they were meeting USA standards. On this first vacation trip from Spain we made a late spring trip to our cabin in glorious Colorado. The morning after our arrival, we awakened to a deep snow, but the sun was out and there was a fast melt. We were startled and then enchanted with the thudding of the snow as it fell from the tall evergreens all around us.

The boys tried sledding on a large red plastic disc, but the rocks and trees resulted in too many banged up heads and knees. However, every day offered something new for each of us in this mountain life. Animal tracks were identified and pursued, water needed to be carried from a river that was now roaring with melted snow from the continental divide, part of our spectacular view. Our cabin included one of those antique black and white coal burning stoves that had a container on the side for heating water. I knew my attitudes about life were changing when decision-making at the supermarket in Boulder left me more traumatized than cooking on that stove. We all ate and slept in one cozy room and for a brief vacation the rugged mountain life was beautiful. The novelty of the outhouse, chipmunks underfoot, and fishing for and then cleaning trout turned this into the greatest possible kind of vacation for the boys.

Back in Spain, pregnant Isabel now had a helper named Tonya. She was a tall young woman, perhaps Basque, although Basque women seldom worked in other homes. Now the boys were completely indulged. Tonya couldn't even understand why I wanted the boys to tie their shoelaces. The boys had learned about male power in Bolivia and they twisted Tonya around their little fingers. Tonya did all the nighttime sitting and heavy cleaning while Isabel supervised the kitchen. She even warmed the boys beds at night with a long- handled copper bed-warmer filled with hot coals. Isabel shopped for most food daily on her way to work, managed our nightmare of a gas stove and of course, there was no dishwasher, washer or dryer, and seven were eating daily in our three story house. Thank goodness I had insisted on bringing the biggest new refrigerator Sears made. It became the miracle of the neighborhood. We had many offers before we left but I needed it to continue on with us.

Often times Isabel couldn't carry all her purchases and would request that they send them on with a boy on a bicycle. One day I called the store and requested the merchant to add something to her order and he said he would also include wine. I said, *gracias* but we still had a bottle and he said, "No, Señora, you can't drink that wine, it's 3 days old". He insisted it must be replaced and sent out two bottles with no charge, which only would have cost six pesetas or ten cents each. That fresh wine was far superior to many of the bottled wines of today, but it did turn into vinegar more quickly. The errand boy picked up our bottle which would be added to their vinegar vat. Like Bolivia, not much was wasted in Spain. Once a cobbler in Spain, having fixed one of the boys shoes, refused payment because he hadn't needed to use any extra materials.

Energy was very expensive and used sparingly in Europe. The electric contraption over the bathtub to heat water delivered only three inches of warm water to our legged bathtub and was cold in five minutes. For a real bath, hot water had to be carried from the kitchen in buckets through the twenty-five foot square hallway, up a grand staircase and down a long hall. The boys shared a bath two times a week, although when Paul jumped into a mud pit at a well site he got one all by himself. Eric and I only got one and I got in first. My regular purchases downtown included light colognes which were decanted into our bottles and splashed on regularly. Not like daily showers in the USA, but sufficient.

Paul samples the mud pit at Antezano I.

The new teacher for the boys' schooling, that now included Paul in the first grade, was a rather stuffy Englishman who seemed to have it in for our independent Pablito. We had given much attention to this shy middle child and Paul had had his way for some time. He decided Paul was 'bright but lazy' and was determined to turn him around but John was 'very bright'. John just considered him 'okay'; Bill was a hard act to follow. The two older sons of our manager, Steve, often fought with our boys, and eventually Eric and I decided that for the next year, the boys would do better to interact in sports and academics with their Spanish friends at the nearby school of the Catholic Marianistas.

That fall we decided to make a trip by ourselves and headed up to France and wound our way through the countryside to Carcassonne. It is truly an amazing sight to see an ancient city and castle within enormous walls, and with a history no less remarkable. We drove on into Italy, spent a night in Genoa and visited the Leaning Tower of Pisa. Then Rome, a museum in itself. The museums were on strike but it didn't matter with such everyday sights as the Colosseum, Forum and Vatican. Rome is such a romantic city; the gourmet restaurants, music, beautiful clothes, plus charming men. Some Spaniards in northern Spain are fair and blonde and so we passed ourselves as Spaniards and spoke only Spanish in public, to save us from the attention that moneyed Americans usually received.

We drove north, spent two days in glorious Florence and then onto Lake Magiori for a night. We left Italy early one drizzly, freezing morning for Switzerland. A dozen of us had to load our cars onto a train for a one hour ride through the Simplon tunnel, but before leaving, the friendly Italians passed out glasses of Grappa and strong black coffee to help us along the long tunnel ride. Potent stuff on an empty stomach and we felt like we just floated through. We were running out of time and money so our trip through Switzerland and stop in Paris were short.

Money was scarce in Europe, not just Spain. Marriage came quite late in life, by American standards, and life was considered a real partnership with all the families as well. Sexual experimentation was common, but young people lived at home, honored their parents, and knew how to avoid pregnancy. There was great respect for the family unit and dignity was an important aspect in personal relationships. Children were loved, indulged and cared for by all family members and if alone, the oldest people lived with their family. In Vitoria, it seemed the only Europeans with whom we could truly connect, were the French. The Spanish and Germans were always reserved with us Americans, for different reasons; we were never invited into a German home.

There were three French families in our neighborhood and several others in town with whom we socialized, all employed in the oil business. They introduced us to their family life, their history and philosophy without hesitation. I studied French at the Alliance Francais, but our friends wanted to practice their English or Spanish with us, so we had little opportunity to practice French. I was told by a French host one night, in Spanish, that I would never be able to understand philosophy unless I studied Greek! Morning coffees with wives and our children were always enlightening. I well remember some of the conversations with these young, yet wise, French women, who considered their families first and husband's fidelity second. Yet they were treated with great respect and dignity by their spouses. These stylish French women believed food was an art, not a necessity. They also believed every child was unique and needed to find their own place in life with little guidance from parents. Money was nice but not paramount in their lives.

An American couple who were in town for a short stay persuaded us to go with them to Pamplona for the running of the bulls. With no overnight accommodations available, we left prepared to sleep in the car and I can honestly say it was the wildest night of my life. We arrived hoping to eat supper but even finding a place to sit down, much less eat, was next to impossible. People from every imaginable country in the world were guzzling *tinto* from '*botas*' (leather wine skins) or laying in the street, drunk. We found a table outside but before we could even find a waiter, a young man came over, introduced himself in some language, laid his head on the table and began snoring.

The event attracted not just European daredevils but the glitzy rich as well. This was the year Hemingway committed suicide and perhaps some were there to pay him homage. None of us got more than a couple of hours sleep that night and then we positioned ourselves along the street early, as advised, for the running of the bulls. It's a frightening event and as usual, a few young men were gored. That afternoon there was one bad kill in the bullring. The people were furious, threw their pillows into the ring, and it took an hour before the spectacle could continue.

At home, we had visits from young travelers who had heard about us and just appeared on our doorstep. Once two American Jesuit seminarians knocked on our door asking for a dry place to spend the night and please, oh please, could they have a peanut butter and jelly sandwich? A nephew of mine, Stan, who was studying French and picking grapes for the summer in France, showed up for two weeks with grape-stained clothes. He had hitched down, a common way of travel for Americans in the sixties. It was safe during that time, even girls bicycled all over Europe.

Our house could be a busy place with delivery boys, the Nuns popping in, the three servants coming and going, my guitar teacher, and boys everywhere. The boys loved any American contacts but also felt completely at home with Spaniards. During the summers, I would drive north with them to spend a week in the fishing village of Lequetio, on the beach near Bilbao. We stayed at a local inn with an enormous and always crowded dining room, like a chautauqua. Breakfast was hot chocolate or *cafe con leche* served with bread from the night before, now toasted and served with delicious homemade jams. The main meal served at two o'clock was simple fare but never-ending and the atmosphere with guests and kitchen help always friendly. The boys loved it all, swimming, digging, building in the sand and making new friends.

Isabel had a little boy, much to the thrill of the whole family. Her mother lived with them, a common practice in Spain, and with two older sisters, I think the baby boy was just loved to death. Husband Felix was beside himself with joy and I'm sure this baby was hugged and kissed by every friend and relative in town and finally succumbed to a virus. The family was devastated. The doctors told Isabel that she could never bear a child again, that her uterus was as thin as paper and another pregnancy would be very dangerous. Isabel and Felix had known this but wanted that third child. She returned to work for us and Felix got great pleasure being around our boys. Felix, like Ricardo, loved taking our John fishing and hunting, bringing home eels and birds to sell to the restaurants in town.

25. Eric: Travels in Spain

There were two geophysicists on loan to CIEPSA and both from London. Doc Gealy, who was approaching retirement, was the Gulf London geophysicist and a delightful human being. Libby and the boys enjoyed his friendship as much as I did and he became a traveling companion and friend to the boys as well. He was a scientist with a sense of humor and although I never had to do much work with him, our friendship lasted for many years. Malcolm Jenyion was a British subject from a Geophysical company that was working with Art to untie some geophysical knots. My first in-depth conversation with Malcolm was at the time of the Cuban Missile Crisis, in a conversation that he requested. He was distraught and fearful that the crisis would escalate, that the United States was acting dangerously and this would evolve into World War III. I could certainly understand his anxiety since he had survived the blitz in London and I tried to convince him that our leaders in the States were not a bunch of wild cowboys, which was one European opinion of us at the time.

I had a similar experience with our gardener, Ricardo. He was from Tarragona where there had been heavy fighting during the Spanish Civil War. His father, who was suspected of being on the 'wrong' side, was dragged out of his house and shot before him. Ricardo fled to the north and made a new life for himself in Basque country. This big strong man became tearful at the prospect of another war in Spain and knew he would be conscripted into any war of ours because we were allies. I tried to reassure him that the American bases in Spain were here to deter war and not to make war. These conversations about war were rare because the people in Europe wanted to forget all wars. Despite living in a police state since their Civil War ended in 1939, the Spaniards were grateful that Franco had kept their country at peace during World War II. They also appreciated the fact that he kept the wine, bread, sausage, and movies affordable.

As Libby and I learned about Spain, our appreciation of European history grew. I began to understand that the Spaniards and Germans had similarities, with their reverence for tradition and respect for order. This made it easier for them to work together but consequently more difficult for us freethinking Americans to fit into this organization that had been built in the previous decade. We made several trips to Madrid during our stay and did a great deal of sight-seeing. The German influence during and since World War II was obvious with the many excellent German restaurants in Madrid. Compared with our food experiences in Bolivia, this European capital with its wide variety of good cuisine made our trips there memorable. The exchange rate was very favorable and allowed us to experience the best that Madrid had to offer.

Apart from El Prado which we visited many times, we also saw the King's Castle-Residence which was too cold and enormous to appreciate. Northwest of Madrid, we stopped at El Escorial, the palace and burial place for the royalty that followed Ferdinand and Isabella. This was the residence of Carlos Primero, who was also Carlos V of the Holy Roman Empire. The enormous wealth that poured into Spain from the colonies in the Americas during the Sixteenth Century made Spain the wealthiest and strongest country in Europe. The Catholic Church wanted to expand its influence in Northern Europe and the church wanted to stop the growth of Protestantism there. Many of the wars to expand Spanish influence were planned in this cold pile of stone. Wherever we traveled we saw remains from the past of Roman bridges, aqueducts and amphitheaters, Moorish Forts, remnants of castles on the plains, the cave drawings, the leftovers of the many peoples who had passed before us.

Roman amphitheater in Segunto.

We were impressed with the Valle de Los Caidos (Valley of the Fallen), which was Franco's Memorial to the dead of the Civil War. It was carved out of a granite mountain, an enormous and beautiful, but somehow spiritless, memorial. On one trip with the boys, we traveled south to Toledo, which had a history that dated back to prehistoric times. The Visogoths had made Toledo their center to unify Spain and where Roman Catholicism was established. After the Moorish invasion in early eighth century, it became the capital of the various principalities that controlled the central and northern parts of Spain. Toledo had been the center of learning in Spain that combined the Christian, Moorish and Jewish cultures. It was a natural fortress built on the bend of the Tajo River and the site of an epic battle during the Civil War that held out against Franco for a long period.

The topography of plains and mountains determined where different people and their language groups settled. Spain is separated from Europe by the Pyrenees Mountains in the northeast, with the Bay of Biscay on the north, Portugal and the cold, blustery Atlantic Ocean on the west, the Straights of Gibraltar with Africa just eight miles to the south and the warm Mediterranean Sea to the east. No wonder there are still four or five languages commonly spoken in parts of Spain. The geology and geography created a crossroads between western Europe and northern Africa. Since prehistoric times, it has allowed many different cultures, languages and races to commingle. In fact, near Burgos, there are evidences of pre-hominid habitation. My work and our life was enriched by the enormous span of archeological history that was influenced in part by the geological foundation of the country.

Gulf was a working partner with Esso, the operator of the Parentis Oil Field near Bordeaux, at that time the largest oil field in France. Gulf had also found the Ragusa oil field in Sicily, the first large oil field in Europe to be discovered after World War II. Gulf had been doing geological exploration work in Europe since the 1930s and immediately after the war returned to expand their work at a time when Europe desperately needed energy to rebuild. I visited the operation in France at Parentis to get information on the kind of formations that were productive in France. I discovered that they had more of the sandy facies that produced oil than we had in Spain and probably better source rocks.

I had to make a number of trips to France. Along with working trips, Libby and I had to take the Peugeot to France every six months to validate the import license. This was a great opportunity for us to acquaint ourselves with the historic cities of Bayonne and Biarritz, the playground of European royalty, as well as practice our French. One of the most striking differences between the two countries was the living style. The Spanish Basque had their *paseo* every evening after work before their very late supper while the French Basque were at home eating their main meal and watching TV, just as we do in the States. Numerous wars had been fought between these two countries and they shared a mutual disdain for each other. A perfect example of this was that the railroad gauge between the two countries was changed at the border, so that troop trains could not enter either country without changing trains. Now it creates a mad dash for tourists carrying their baggage across the border and over the tracks for the next train out of Irun, Spain or Hendaye, France.

One day I was contacted by the city authorities who asked me to come to the police station to translate for a young American. They explained that he spoke no Spanish, had been in jail for several days and needed help. They did not want me to tell him that several young people had been killed as a consequence of his driving. They felt he was too young to absorb such a tragedy or to be prosecuted. I was told that there would have to be proceedings against him to settle insurance claims and they only wanted me to tell him that there were no serious injuries. They brought a very frightened David into an interviewing room and I began to translate for the police. They wanted me to tell him that he would have to be in custody at the police station or under house arrest in our home until the trial, which could be as long as two months away. I explained all of this to David and invited him to stay with us, which relieved him considerably.

This was an example of Spanish compassion and their consideration for foreigners. It was the second time we had been contacted to intercede for Americans passing through town. The locals obviously felt we could be trusted to handle these situations appropriately. The New Yorkers in the hospital with concussion and broken arm were also frightened and I had to reassure them that the Spaniards would do the right thing. So David was not our first encounter with Americans in trouble. We never knew whether the locals just had confidence in us or whether Rafael, the top Spaniard in the company, had fingered us for these volunteer jobs. Whatever, it made us feel part of the community.

26. Libby: Travels with the Boys, and David

In July 1962 we began traveling home for our summer vacation with a trip through France. While in Paris we watched the famous, tall, and ramrod straight President De Gaulle march in the Bastille Day parade. Their very tragic Algerian war of almost two decades was lost and the French Foreign Legion was being disbanded. Tears flowed everywhere. Although we were unable to understand the tragedy, their pain was palpable. It was a memorable experience for us and all the pomp fascinated the boys. We had a few days for viewing paintings at the museums, sculptures and historical buildings; we floated on the Seine and climbed the Eiffel Tower. We visited an unforgettable cheese shop with counters and shelves that displayed only different kinds of cheese. Eric, a good Swede, was ecstatic. Crowds of people were sampling slivers of cheese sliced by white capped servers behind high counters. I believe it was De Gaulle who said that it was impossible to govern a people that produced so many different kinds of cheese.

We next boarded a train for London, traveling first class. Very proper English waiters served us 'high-tea' with every imaginable type of cake, cookie and even the petite cucumber sandwich, served on the finest china and hand embroidered linens. In London we visited museums and saw the crown jewels but the boys favorite, of course, was the Tower of London, with its armor and very bloody history.

We had scarcely settled back home from our stateside vacation when there was another car accident in Vitoria with an American, this one quite tragic and with sad consequences. A young American named David, only seventeen, had not been admitted to the university of his choice and to compensate for his disappointment, his parents gave him the money to buy a new Volkswagen in Germany and to travel through

Europe. He had already driven through Vitoria and was on his way to Burgos when he realized that he had left a gold pen, a graduation gift, in his room of the night before. His parents had told him that he could only drive during the day, but he decided to return that evening to remain on schedule to meet friends. He drove into the outskirts of Vitoria on a Sunday evening and said in court that he was blinded by truck lights. He plowed into a group of teenagers, killing three.

After being called to the police station to translate for David, Eric brought this emotionally disturbed young man home to us to stay indefinitely. These compassionate policemen and magistrates thought he was too young at this time to know all that had happened. Isabel and I settled him in the bathroom for a much needed bath and went downstairs to work in the kitchen. We heard the water drain from the tub and soon the pacing began. He walked back and forth in the corridor until I went upstairs. I was no psychologist but it was obvious that he was very disturbed. The best thing that could have happened to him was the boys. When they came home from school, they all began to wrestle as boys always seem to do, with Loompi joining in.

We were in touch with the American consulate in Bilbao, and with David's family on the phone, immediately after his arrival. In rare cross-continent phone conversations for those days, we assured them that we would listen to him, take care of his needs and keep them informed. When Eric and I were both gone one day, the consulate from Bilbao called and told David that he had killed three young people. We were furious when we returned and discovered what had happened. The Spaniards had sensed his youthful vulnerability and despite how they hated the act, they had reacted as we did and wanted to protect him. Now David was in real shock. He did not want any family member or friend to come for the trial. A girlfriend who was traveling in Europe joined up with an aunt of his and visited, but he would barely speak to them.

The Consulate took over at the trial and neither Eric nor I were requested to be there. David was an American, driving a new German car in Spain, with German insurance that had to pay the injured parties before his release. They accepted his word that he was blinded by lights. He was taken to the border where they returned his passport and car but he was forbidden to ever drive in Spain again. It was also decided that the families of the three teenagers would receive two thousand, five hundred dollars each. That was the worth of a young life in Spain in those days. After the trial, grandparents of the victims came by our house in tears to tell us that these grandchildren had been their insurance for old age. It was a miserable time. David decided he would not return home but would study in Paris for a year.

This last school year in Spain, John and Paul were thrilled to begin school with their many Spanish friends. The padres gave us the option for them to either attend catechism classes, or not, and we discussed it with them. Fourth grader John decided 'no' but second grader Paul wanted to be with his friends all the way. Our boys were always good athletes and with their American spirit they were subject to some hero worship. But now their friends cried and told them they would go to hell if they didn't join the Catholic Church and John couldn't figure out why his many Catholic friends in Bolivia hadn't worried about him before. We decided to emphasize respectful behavior and just observe. Franco did not allow Protestant churches in Spain.

After Paul's experience with the French kindergarten, I decided to begin a kindergarten of sorts for at least three days a week. Besides Mark, this would also include the little girl of my American friend, Glenda Chapelle, and her French husband René, and the daughter of our Geophysical department head, Art, and his wife, Lisa. The wee ones spoke to the servants in Spanish, to me in English and in French, Italian and English between themselves. They had a big playroom upstairs with large glass doors overlooking the pastoral scenes of the convent across the street. The dear owners of our house, Jose and Blanca Alvarez, did not even object to the kids roller-skating on the third floor during rainy weather.

Scott Carpenter followed John Glenn into outer space, became entranced with the view, delayed pushing the right buttons for a second or two and as a result had to float around in the ocean in his capsule for about an hour. It was a dreadful wait for the world and for us. Eric and I laid on the floor with our radio in the best position for reception, waiting for an interminable amount of time. I was convinced that Rene would have four children to raise on her own but miracles do occur and they not only found his capsule but managed to rescue Scott.

December arrived and another memorable European Christmas was approaching. With the stores closed from 1:00 until 4:30 PM, much shopping was done later in the dark. The tiny stores were often candlelit and smelled of coffee beans, candied fruits and kerosene or gas stoves, and all the street corners had charcoal stoves with roasting chestnuts. There were colorful hand-knitted mittens and scarves everywhere, some toys, and a few decorations could be found. Turkeys, ducks, hens, pheasants, rabbits and other small creatures were hanging in the windows of every fowl shop. These nights may have been snowy and cold but the proud Spanish store owners were so friendly that chances were that you would also be offered a glass of sherry to cheer your spirits on a cold night.

Late one evening, several days before Christmas, Loompi kept barking at the front door and alerted us. Finally we heard a quiet knock. We were stunned to find a disheveled, cold and quiet David standing there. He had hitchhiked into Spain to spend Christmas with us. John heard all the commotion, peered over the staircase, saw David and raced downstairs to throw himself on David's back. When we were alone in the living room the next day, David told me that he now agreed with the Nazis: that some people just didn't deserve to live, etc. I was dumbfounded. I took him downtown one day to pick up our turkey; he saw someone on crutches and under his breath muttered "and I didn't even have a broken leg".

Days before, I had arranged at the fowl shop to save me the biggest turkey they had. I had also asked specifically that they clean the turkey before hanging it in the refrigerated room of the hotel. I pushed through all the hanging animals in the shop (no joke!), waited in a long line and was finally taken aside, told that our twelve pound turkey was beautiful, and 'wait and we will clean it'. I was dismayed but was finally beginning to understand my place in this country. Being a simple American, I just didn't appreciate that all fowl tasted better having hung with its entrails for at least three days. Isabel was offended when I gagged after smelling our beautiful turkey but I noticed neither guests nor family wanted to partake of the lower end.

After spending Christmas with us, during which time we tried in vain to persuade David to return to the States, he hitchhiked back to France where he completed his year at an American College in Paris before returning to the States. We visited his family once in Oklahoma while on vacation and then saw him for the last time in the early 1980s in San Francisco. He was a married lawyer with children. It had been a terrible tragedy and a growing-up experience for all of us. The boys only remembered the fun they had with David, for which we were always grateful.

For some time we had wanted to travel south to Granada and Cordova and after New Years, the time was right for both Eric and the boys. Driving south from Madrid, we were surprised to see the Guardia Civil standing at attention with rifles lining the highway every half kilometer for miles and miles. When we were finally stopped, we asked one what was happening and he shouted, while still standing at attention with his rifle, something like, *"El Caudillo viene a cazar!"* (Franco comes to hunt!). We definitely got the impression that we shouldn't have asked. Soon Franco and his entourage passed by with as much protection as you would see an American president travel today. All the cars had black-curtained windows and they traveled fast. Franco's control was total in Spain and had been for thirty years.

On that trip we toured the Moorish Granada, the delicately beautiful Alhambra, even its caves with dancing gypsies. Then on to historic Cordova with its enormous Mosque built with Roman columns. Carlos the First had forbid the people to tear it down, and so, instead, they built a small Cathedral within. The history of Catholics, Jews and Muslims, how they mixed and survived each other through the centuries in Spain, is fascinating stuff. It was a very memorable trip with the boys, despite the rain and cold. John was now ten, Paul seven and Mark four. They had become curious and seasoned travelers while we had been in Spain.

While staying in the Canciller Ayala Hotel on our arrival in Spain, we had heard Andres Segovia practice his guitar in his room and were informed that he did so eight hours a day. We were eager to hear this famous guitarist and our opportunity arrived one winter night in Vitoria. However, before he played for us that evening, he sat with his guitar and stared at the audience for several minutes until all the coughing and sneezing ceased and there was absolute quiet. You could have heard a pin drop, truly. His command of the stage and audience was memorable, not to mention that of his guitar. We saw and heard him play his beautiful music twice again in the States, but he didn't attempt to display control of the audience like he had in Spain.

Afterwards, we joined friends for desert and coffee in a local bar. I had some strawberries and whipped cream which turned out to be very bad news for me. In three days I was delirious and taken to the hospital with a raging fever that I have thankfully never experienced again. The immediate illness caused me to lose 25 pounds and become weak as a kitten, but I slowly got better. Doctors in Spain and New York City thought it was just a strange virus but I continued to have intermittent fevers with exhaustion.

27 Eric: Vitoria I blows in, on to Nigeria

By the beginning of 1963 we had drilled several additional wells in the vicinity of Castillo I. It was beginning to appear that it would be a one-well field. The geophysical data had been reworked sufficiently so that we could see no further gas or oil prospects in that immediate area. I had also reviewed and compiled subsurface data from the drilling that was completed in other areas before I had arrived. I observed that oil had been generated in some of the limestones of the cores from one of the Trevino wells. From this work I saw some possibilities of another location and suggested this to Ossie. As a result, the company decided to drill another exploratory well in the Lano area. The well, Lano III, just produced water. Most of the porosity had been destroyed by several generations of cementation and fracturing. This was all described by Max in some detail. There just didn't seem to be any oil left in the area. If there had ever been any oil there, it had migrated elsewhere.

The Spanish geologist, Angel Paradinas, had a familiarity with the details of the German well evaluations and felt the work had been thorough. Angel, a friend to this day, had made the transition from mining geologist to oil geologist and I valued his observations and opinions. He also knew of other areas in northern Spain of some dead oil, or bitumen, but no seepage of 'live' oil. It wasn't expressed, but there was a palpable feeling that there really wasn't more that we could do. There was an area about one-hundred miles south near Soria that had been well-known for its accumulation of dead oil, possibly from a dried-up oil field. It had been studied in detail but no production had been found in the vicinity. West of Vitoria, northwest of Burgos, Caltex had just found a small oil field that subsequently produced about a million barrels of oil with the Spanish Government as partner. All of these efforts seemed to have exhausted any possibility of large fields in northern Spain. Other companies drilling adjacent to the CIEPSA permits had focused on salt-domes and had only found shows of dead oil and nothing commercial.

South flank of the Pyrenees in Catalunia

I had begun to understand that this exploration effort in Spain by Gulf Oil Company was secondary in importance. It was an obligation for Gulf to invest money in exploration for oil on the mainland to secure a concession for the Spanish Sahara (Mauritania), which was their primary interest. The decolonization of Africa forced Spain, France, Italy, Portugal, and Britain to evaluate vast areas of their colonies before they would become independent. Most of the Gulf geologists in Europe were being considered for the African exploration programs that would be developed. However, I did not want to give up on the shale gas potential in Spain. I thought we could find a better prospect area than the Castillo and thought I knew where it would be. Nobody in the Gulf organization was interested in gas production but the Spanish needed more energy and there would be a ready market, even in small amounts. Despite their belief that there was little potential for large accumulations of gas, Gulf was obliged to do the exploration work on the mainland.

I had heard about air-drilling on our last trip to the States and I wanted to experiment in an area south of Vitoria. Ossie supported me, which was important, because Diehlman would not be able to use their rigs. Gulf then agreed that drilling a fractured reservoir with air would give a continuous test that would not be damaged by using a heavy water-based mud. The Gulf Drilling Department in Pittsburgh found that a Denver Drilling Company had an air-rig in France and was able to bring that rig to the location south of Vitoria. The surface geology at this location suggested that there would be increased fracturing at depth, perhaps more than in Castillo. It was on the plunge of a regional anticline to the west and it would seem to offer greater reservoir possibilities. After the surface casing was set, we began drilling with air in the fall.

The regular semiannual Operator's Committee Meeting with representatives of the head offices of Gulf, Diehlman and CIEPSA took place three weeks later. On that night, the well blew in with an estimated flow of about five million cubic feet a day, of almost pure methane gas. The noise awakened the neighborhoods for several miles around and the visitors all went out that night to view the monstrous flare. Garland had left for Nigeria and had been replaced by Tubby Knight, an American, who had had considerable experience in drilling completion in the States. The drilling on Vitoria I was continued but the gas began to diminish with our hopes. The casing was set and the well completed but the gas production was minimal. We suspected that the well had been damaged in completion. However, a second well drilled nearby confirmed the meager results of Vitoria I.

During the drilling of Vitoria I, I received a letter from the head of the Spanish Gulf office in Madrid offering me a job in the new exploration effort in Nigeria. Initially I was not enthusiastic, despite the exciting talk among Gulf management. After previous disappointments in Libya, Tunisia, Ghana, Angola and Mozambique, this was the best opportunity for Gulf to get production in Africa. Despite the potential of opportunity in a large production, I had hoped there would be more exploration to keep us longer in Spain. I was worried about the undiagnosed illness from which Libby was suffering and the boys were so happy in Spain. However, my responsibilities had essentially ended with CIEPSA after Vitoria I, Steve had been transferred to the Canary Islands, Garland had just arrived in Nigeria and Art was already in Somalia. In Nigeria, I would be number two geologist with an uncertain future in a new organization. But if we found big oil, and the prospects were good, I would be on the ground floor. Finally, since Libby hadn't had her recurrent fever for some time, we decided to take on one more foreign assignment.

Looking back on my three years in Spain, I felt that I had made a contribution to the effort to establish their first gas production. Subsequently Castillo I produced a little over two billion cubic feet of natural gas which furnished energy for several factories in Vitoria. However, I knew my work in Spain was simply fulfilling an agreement with the Spanish government and economic results had not been expected. My work had been appreciated by the Spanish and Germans and I had formed lasting friendships with both Americans and Europeans. In these three years, I almost thought I had achieved a Ph.D. in European studies. I had worked with the poor, the rich, as well as some aristocrats.

I also felt I had achieved an advanced degree in company politics. There were many company visits from the States and the Gulf London office, during our three year stay. Members of the Gulf executive staff were eager to stop in beautiful Spain on their journeys to other Gulf operations. Many of these men had worked in Latin America, spoke Spanish, and wanted to visit the 'homeland' and it was also a learning experience for them to travel in Europe. I was beginning to understand that those from the South American operations were climbing the executive ladder and were in competition with others who had experience in Europe and the Middle-East. Gus, who had originally hired me for Bolivia from the New York office, had been in charge of exploration in the Eastern Hemisphere for a few years at the London office and was now scheduled to return to Pittsburgh. He was in that group of visitors when Vitoria I blew in. Now all of these men were wondering how they would fit into this new developing Africa, including me.

Among other things, I learned that the Kuwait Oil Company (Gulf-BP) had replaced Venezuela as the largest producer of crude oil for a rapidly growing demand in Europe and the Far East. Gulf Oil Company had grown from a single discovery in Texas to become one of the seven largest oil producers in the world. Now it was important to have international experience for any advancement. There was strong competition between the heads of the regional areas of Gulf to get to the top spots in Pittsburgh, the head office. Now I was part of the international chain and at thirty-four, I was becoming a competitor. I didn't enjoy administrative responsibilities that much but I knew Africa would help in my resume, no matter what my future.

Ed Walker had come to London from Venezuela where he was an executive with Mene Grande Oil Company (Gulf). He had replaced Gus and on my way to Nigeria, I was scheduled for several days there to discuss Africa. Ed wanted to find out all about Spain, my exploration experiences in the States and discuss our mutual ignorance of geology in Sub-Sahara Africa. He was personable and smart and I enjoyed a dinner with him and his wife. Eventually he would become CEO of Gulf Oil Corporation. I flew out the next morning from Heathrow on BOAC on a direct flight to Lagos, Nigeria. By chance, I met my two future next-door neighbors, Jake Jacobsmeyer and Darryl Hatchenberg, excited newcomers to foreign life on their first trip out of the States. Both were geophysicists, would be working in a new Gulf geophysical and geological regional exploration office, that would be reporting to Gulf Research and Development (GR&DC) in Pittsburgh. It was sheer coincidence that we had met since their flight had originated in Pittsburgh.

I was much less enthusiastic about this new adventure to Africa than my two traveling companions. I had grown to love Spain and was loathe to leave. However, there was the thrill of going into a new operation at the very beginning in this newly independent African nation, Nigeria. In December of 1960, the Nigerians had been given their independence from Great Britain, after one-hundred years of colonial control.

Nigeria was the most populous African country south of the Sahara and I knew that there would be many exciting challenges in my work. But my anticipation of the business opportunities was tempered with my concern for the family and how all of us would adjust in this new environment.

28. Libby: Isabel, Joan, and Marianistas

Isabel got pregnant again, much to our dismay. Perhaps she and Feliz decided that they must try again for that third child. She continued working until she was two months short of delivery and then Tonya and I managed quite well as I seemed to have begun to recover from my periodic fevers and weaknesses. Then we got word that the doctors refused to deliver Isabel's baby by Caesarian section when she was eight and a half months pregnant; that it was not in the best interest of the baby and Isabel be damned! Finally the word came that her uterus had split open with another baby boy and Isabel had bled to death. It is impossible to express my absolute fury. I just couldn't believe that the Catholic Church would cast her aside in such a barbaric way. We kept it from the boys for many years, telling them that she needed to care for the baby but they always wondered why she didn't visit us.

Our life in these Catholic countries had resulted in mixed emotions for both Eric and me. I looked across the street at the nuns going about their daily caring services in the community. When I was sick in bed, they not only brought flowers and sweets, and played the guitar for me but they always took Mark, joyfully, when needed. We saw John and Paul come home happy and challenged from the Catholic school, receiving the praise and support necessary for their mental and physical growth in a foreign country. We saw and felt the compassion of the community for David's tragedy. I remember well my hospital stay and how an older male nurse had held me in the middle of the night as I shook with delirium and that high fever, telling me over and over that now I was going to be all right. How could these caring, friendly people have condoned what happened to Isabel in the name of religion? I still find it difficult to believe. I did, however, albeit for different reasons, begin to understand why the Basque were often opposed to the Catholic Church, as well as Franco. Franco and a few religious leaders had absolute power over the people.

Eric had a business trip through Europe and his sister, Joan, had flown to Stockholm to vacation with cousins and she and Eric arranged to meet in Paris. They had adjoining rooms but she had no bathroom and so, of course, she shared Eric's. She was a beautiful 5' 10" blonde Swedish type and they enjoyed the many smiles of his associates when he introduced her as his sister. She asked Eric what the sink-like-toilet was in his bathroom and he shamelessly said it was for soaking socks and underwear and it was called a *bidet*. She thought it was cute and toured the jewelry shops in Paris searching for miniature gold bidets for hers and my charm bracelets. She traveled on down to Vitoria and proudly presented the gold bidet charm to me. When I explained her brother's joke, she understood why the jewelers kept asking her with a surprised voice "A bidet?".

We went back to the States for an early spring vacation and the doctors in NYC put me through a complete physical and a thorough battery of blood tests. They found nothing and I was beginning to question my sanity. They assumed I'd contracted some kind of virus that had now passed. So Eric and I considered all our options. We discussed going to Africa, leaving the Company and returning home, or possibly even staying in Spain. When we broached the subject of leaving with the boys they were raring to go, although they loved Spain. I think they had already learned, as we had, that it was not difficult to make new friends, or perhaps they had just inherited our 'itchy feet'. Ultimately, the lure of Africa won, especially since I had been feeling good for some time when we made our decision. Then, after all, could any place be more difficult than Bolivia?

Ricardo told us that Isabel's mother had died and Feliz married a widowed neighbor to take care of the baby boy. Just before we left the house, Isabel's daughters, now 15 and 16, visited one afternoon. They were small and rather meek girls and the stepmother, whom they had known for years and did not like, insisted they move out of their home. Both were now working in a factory and their employer was making passes at them. They had heard we were leaving and through tears, begged to go with us. They didn't want money, just protection. Of course, we couldn't take them and I could only insist that they go to the church for help. Besides my illness, we had suffered the tragedies of David and Isabel, and now, her daughters. They broke my heart.

Eric and I watched our things being packed and could only keep our fingers crossed for the long voyage to Africa. Eric would precede me by two months, select our house from those that the company had leased, and get started in his exploration. During this period I would be back in Jose's apartment-office building and have time to buy new clothes for this dramatic climate change. The temperature in Lagos averaged 85 degrees all year with humidity hovering around 100%. I bought shorts and the coolest shirts I could find for the boys and had a seamstress, who had trained in Paris, make me a few bare dresses of cotton and silk that were a godsend when we arrived in Lagos. I had worn mostly wool in Spain and dress for both men and women was quite formal in those days. Hats and white gloves had essentially been chucked by women, but for us, slacks outside the home were still a dream. Not knowing what the future held, I packed up the woolens with mothballs in trunks and sent them off to Africa.

I had never been inside the Catholic Parochial school of the Marianistas. Given it was a boy's school, Eric registered John and Paul and had been the go-between when necessary. However, just before the end of the school year and knowing we would soon be leaving, I wanted to thank the padres who had been so helpful. I walked the few blocks to the school and began up the steep steps when little boys, playing outside, ran up, asked if I was John and Paul's mother, and then escorted me inside. A padre soon appeared and after I introduced myself and explained my mission, he had me sit in a waiting room and rustled off. I waited and waited, was becoming irritated and even thinking about leaving, when I heard more rustling and several padres entered the room.

They introduced themselves and I thanked them individually, and collectively, for all they had done to help the boys adjust to a strange and foreign situation. At this point, their spokesman offered a little speech. He said, in essence, that they were the ones that needed to thank us for having two such special boys and bringing them to their school. He said that previously they had not known Protestants and would be forever in our debt for enlightening them! I was almost speechless but managed to find a couple of those rich Spanish phrases of thanks and farewell. They each bowed over my hand and I took my leave. The dichotomy of my thinking concerning the Catholic Church was not lost on me.

It had been arranged that one of the CIEPSA chauffeurs would drive the boys, Loompi and me, with all our luggage, to the French border and put us on an overnight train for Cannes, France. We would arrive in the morning and depart from the airport for Lagos in the early evening. The thought of a day in the luxurious city of Cannes with its famous beach had sounded wonderful to us, but during the night on the train, my old enemy, the fever, returned. I could barely take charge the next morning and had to lean on big brother John to keep Paul and Mark under control. We checked our baggage at the airport, took off for the beach and despite our traveling clothes, the boys and Loompi managed to chase each other around in the sand and expend some energy.

While looking for a restaurant for lunch, two chefs ran out and competed with cutup steak and French baby-talk for Loompi's attention. The French and English seem to dote on dogs even more than Americans. We had to put Loompi in a cage and bid him farewell before we boarded the plane. He probably remembered those long flights from Bolivia to the States and then later to Spain from the States. Goodbyes were difficult to that loving bundle of muscle and energy.

As we watched the sights of Europe fade in the distance, I wished that all Americans could know more of their ancestry and learn to appreciate the generations that had brought them to this time and place in their lives. After this stay in Europe, I felt I had made inroads into my ancestral history, not by name but by association, history, and a feeling of belonging in certain places. I did look forward to reading the news again, having lived in a desert of information where Franco would not even permit Time magazine to be sold. However, this man and his regime had given us a physically secure life that enabled us, as visitors, to explore new cultures. In Europe we had been exposed to a gentleness, a quietness, if you will, a new depth of understanding for other peoples and an introduction into other philosophies. We had begun to think of Europe as 'the old man' and the USA, 'a teenager'.

We knew we would return to Spain and Europe, it was already on our agenda. On the flight, I only remember watching the boys sharing this experience, so close together, so happy and excited about seeing their dad again. There was animated talk of natives, wild animals and English speaking schools. As for me, I didn't see how I could survive a stay in this new continent with this unidentified illness. However, I knew these fevers with weakness and depression that I had been left with, after the initial attack, had always passed before, and I began to think only of our family reunion. Hadn't I conquered all those illnesses in Bolivia? Yes, I was a born optimist with a sense of adventure.

Part III
NIGERIA
1963–1966

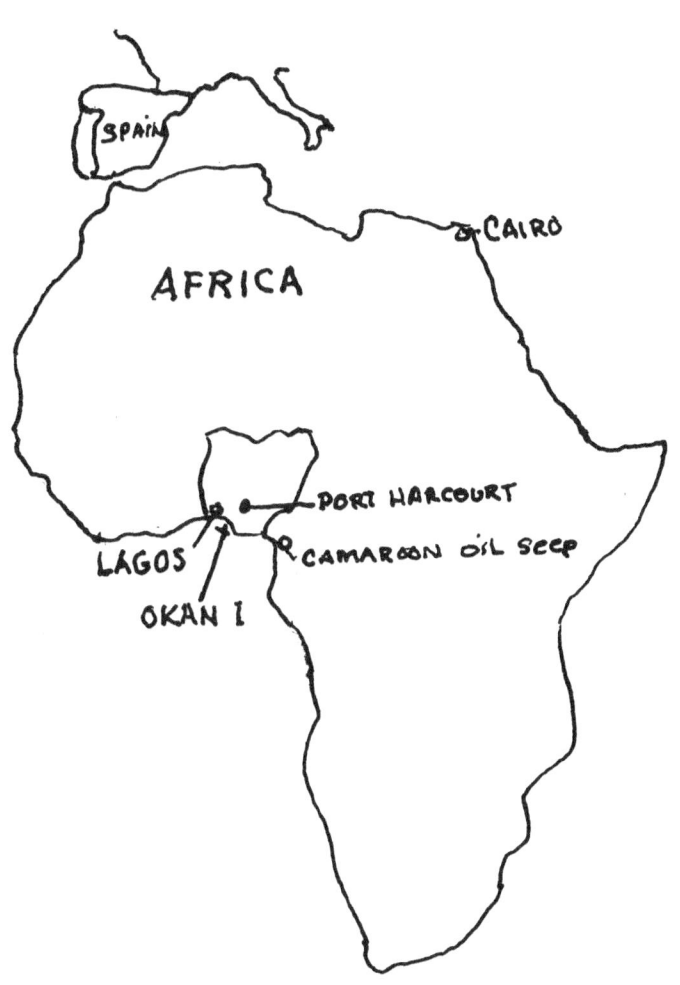

29 Eric: Steamy Lagos

Jake, Darryl and I marveled over the extent of the Sahara Desert on our flight to Lagos. As we descended and the cabin was being depressurized, we became aware of the humidity that we were going to be living in. We were met at the airport by Doc who was his usual jovial self. We went through customs and immigration quickly. It was a new experience for all of us to be passed through in an all-black officialdom wearing smart uniforms. My first impression was that they enjoyed being in charge and were quite proud of their new country. I was decidedly glad that I had bought tropical clothing while in London but even that seemed heavy in 85 degree temperatures and what felt like 100 % humidity. We were only a few degrees north of the Equator and although it was early evening, the sun had already dropped. The drive was about forty-five minutes and the road was lined with small stands where vendors, with little candles on their tables, were selling all kinds of items.

Rooms had been reserved for us in what appeared to be a downtown commercial type hotel. The rooms were sparsely furnished but clean. The heavy traffic was noisy and chaotic outside our windows, the air-conditioning was barely adequate, and we didn't sleep well. The next morning when we left our rooms and met for breakfast, we had to pass through an open atrium and were almost overcome with the odor of the city, which smelled like soiled wet diapers. The humidity and smells were almost a physical sensation. In the dining room there was a buffet breakfast which consisted of boxes of dried cereal, cooked oatmeal, strange cutup fruits, limp bacon and grey scrambled eggs. There were also the British kippers, which delighted me but proved to be quite unlike the Scandinavian herring I was accustomed to. All of us immediately shared our anticipation for assigned housing, with our wives managing the kitchen.

Nigerian officials.

Doc picked us up and drove to the temporary offices of Nigerian Gulf Oil Company downtown near Tinabu Square, above the Barklay Bank Building. We met our new manager, Glenn Ledingham, who had worked for the Company for thirty years, the senior geologist, Gene Cordry, from Mene Grande, both Americans, and Charlie Smith, the Scottish accountant from Gulf's London office. It was obvious that this was a beginning operation although all of us had had at least ten years of experience. Doc would be our geophysical advisor until Price Barbour, to whom Jake and Darryl would report, would arrive. Garland, who had recently left Spain, was in the field organizing a temporary shore base for the offshore operations. He would be the drilling engineer for our first wildcat which would be about seven miles off the coast. Gene said the rig was in Nigeria but would be shared with Caltex, and first, Caltex would have the opportunity to drill one or two wells on their exploration license.

Since Caltex would have the rig for a couple of months, this would give me a unique opportunity to study the geology in this part of Africa, unlike my previous experiences in Bolivia and Spain, when I had to hit the ground running. As it turned out, I would have about seven months before drilling operations would begin. Gene said there was very little published material and so I sent off for the French Colonial Geological surveys for the areas adjoining Nigeria. This would supplement that which I would be able to obtain from the British Colonial Survey. Gene and I decided to travel to the British survey office in Kaduna, the capitol of one of the predominantly Muslim provinces in Nigeria, about two-hundred miles north of Lagos.

Gene made reservations for us in the British Guest House in Kaduna where there were reasonable accommodations and European style food. The British system to aid travelers in their former colonies was absolutely necessary in Nigeria. Unlike Bolivia and Spain, it was not possible to find anything similar to European accommodations or food except in five or six large cities. We rented a car, drove through the countryside on dirt roads and saw only hard rock country, which was not of interest for oil exploration. The British were helpful but didn't have a great deal of information. The morning we left we saw two barefooted women dressed in black carrying enormous baskets full of wood that would be difficult for any man to carry. It gave us a living example of what it was like during the time of the 'hunters and gatherers'.

Back in the office, Gene told me that two other geologists, John Recamp and Whitt Stucker, would take over the well-site work and soon would be arriving from the Turkish operation in Ankara which had finished its drilling activities without success. More geologists would be available from Mene Grande, where the Venezuelan government was assuming more control over the industry. If we were successful, we would also be competing with Bolivia for all exploration personnel, because of our new found production there. The Nigeria Shell BP operation, which began shortly after World War II, was finally having success after many years of dry holes. Several geologists with Shell BP, that we had known in Bolivia, were now in Port Harcourt.

The Gulf exploration licenses consisted of two adjoining offshore blocks near the western limits of the Niger River Delta as well as several more licenses in the eastern onshore part of the Delta. The size and orientation was similar to the Mississippi River Delta offshore Louisiana. The drilling operations of Shell BP were managed from offices in Port Harcourt, whose location could be compared with New Orleans on the Mississippi River. They had found the first oil production in the onshore swampy area in 1956 and had conducted a feverish campaign to explore the entire Delta area before Nigeria would become independent from Britain in December 1960. After independence new areas would be awarded other oil companies based on competitive bidding. Gulf had been able to secure what we believed to be choice areas largely because of the interest of two geologists, Hollis Hedberg and Gus. Hollis was vice-president in charge of exploration in Pittsburgh and had been instrumental in securing exploration areas in Mozambique and Angola in Africa, still Portuguese colonies, that were now being explored by Gulf.

For this small office we had hired a Nigerian administrator, Tom Big Harry, who doubled as a personnel man and remained with us for several years. We had also hired a Nigerian secretary, Minneka, a statuesque woman and a member of the Royal Calabari family in southeastern Nigeria. She had been educated in England and both she and Tom spoke the 'King's English'. Tom dressed in western clothes but Minneka in tribal clothes, at times heavily embroidered and very beautiful. Unfortunately, she was only with us a short while because her father died and she had to return home. Her father was the King of Calabar and she was obligated to remain and protect his body for a certain period after his death. We were stunned to find out that body parts of royalty were coveted for their magical powers.

I had worked for Conoco for five years in the interior part of the Texas-Louisiana Gulf coast and one of the reasons I wanted to leave was to experience other geological regions. I had certainly managed to do exactly that in Bolivia and Spain. But now, I was back in the offshore analog of the Gulf coast of Louisiana, except there were no salt domes as far as we knew.

30. Libby: Federal Palace Hotel and the Fever

When we arrived in Lagos, Nigeria, the boys were excited but I was exhausted and apprehensive. Eric and some of his associates had already been there for a couple of months but most were still waiting for their wives and children. Much of the groundwork had been done for the Company, the Nigerian Government wanted us to continue the successful oil exploration that had begun with Shell BP, and houses had been allocated for all the incoming families. Eric would eventually be in charge of the geologists and big things were expected of all of us.

It was 3:00 AM and very dark when we flew into Lagos. The only real lights were at the airport, and except for a few flickering candles around the airport, there was no indication that close to a million people lived in this city. I struggled off the plane with three sleepy boys, one very happy dachshund, and an enormous amount of luggage. The smells and heavy air nearly knocked me over. Humidity is one thing (Houston comes to mind) but mixed with the smell of sewage, it takes on an extra dimension that is difficult to describe.

Eric and our future next door neighbor, Jake, whose family had yet to arrive, were waiting behind a metal gate, not allowed to enter, only able to shout greetings. The boys soon accepted that they couldn't be hugged by their Dad and collapsed on a bench where the youngest two fell asleep. I was immediately addressed as 'Madame', pronounced 'Mahdahm', and I remained Madame for the next three years. When all the Nigerians and regulars had passed through, our luggage was finally opened, and I had no idea what to expect from customs. I had learned in Bolivia about *mordida* (the bite) or bribes, and so was prepared to deal with what was called *dash* in Nigeria. Gulf had been opposed to money bribes since we had begun on our foreign journeys and although I wasn't aware of the exact situation in Nigeria, I assumed things hadn't changed.

These smiling Nigerian men began going through all our luggage and soon found something humorous. It was a hairdryer (a simple affair of those times) and they began showing it around, speaking in what I was to learn was typical broken English "Look what Madame bring! What it for?" I explained and they laughed and joked in their language and kept after me to explain. Finally I said it was my gift to them and they said, "Oh-ho! Madame give us gift!" I wondered what it would be used for.

I had loosened the strings on my guitar for air travel since the altitude could cause the strings to shrink and tighten, cracking the wood. They opened my case and said, "Oh Madame, dis beautiful geetar! Very expensive geetar!" Again much laughing. I smiled and said, "Thank you, I wish it was but it's really not that good, try it." They did, passed it around and the 'thud, thud, thud' they heard convinced them I was right. Fortunately, guitar players were not common in Nigeria. "Oh very true, Madame, very bad geetar", and handed it through. I still cherish this story, particularly with the Nigerian scams that have occurred worldwide since then. I will never know what possessed me to be so clever that dark night in Lagos, Nigeria.

Finally they let the boys fly into their father's arms and we packed into two taxicabs to drive to the Federal Palace Hotel, the only large, modern hotel in Lagos, where we would stay for several weeks. The boys were adventurers personified. If we had had one child or perhaps even two, it would not have been the same. Absolutely nothing seemed to frighten or slow these three down. They had each other and adoring parents besides and these years were wonderful times for Americans in foreign countries. Our towheads seem to charm everyone they met.

I remember little of the next few days. I was not well, the fever did not go away, and I laid in bed. I had come to Africa with much trepidation, given my physical condition, and I told Eric upon arrival that I must see some English doctors now or I would have to go home. Eric was beginning to understand that the problem was bigger than both of us and all the doctors in New York City. He immediately brought two doctors into our rooms, but separately. One young, tall and sensitive, the other much older, pudgy, cantankerous, and both British. Each suspected brucellosis, a totally new diagnosis to us. They insisted on blood samples and it would take weeks to grow cultures. No one had done THAT in Spain or in the States.

 I got better and began to enjoy life in the waiting. We made trips to Tarkwa Beach on motorboat taxis to picnic and swim with other newly arrived families. The boys loved the 80 degree humid climate after years in cool, rainy, northern Spain and were treated like princes in the hotel. As for princes, one night we came down to dinner in the large dining room and I was agog to see the room filled with tall Africans in white naval officers uniforms. They had just disembarked from an Ethiopian ship and most were part of the extended family of Haile Selassie, the President of Ethiopia. During our years in Nigeria, I often heard Nigerians describe other Africans or Nigerians as that 'yellow', 'red' or 'dark' man. We were classified as Europeans and these Ethiopians would probably be called 'yellow' men.

 Our boys, led by their indefatigable leader, John, were constantly on the move and not only watched but encouraged in their every endeavor by all the Nigerian employees. They had learned to pay attention to our guidance in our travels, but there was one unforgettable experience with Paul that we could never have foreseen. The hotel was surrounded with tropical trees and plants and they chased each other through beautiful flowering bushes to run off the energy that they couldn't in the swimming pool. However, curious creature-loving Paul was always crawling around looking for lizards in the bushes and suffered the consequences. He began scratching his head and we noticed bumps. They became larger and inflamed so we took him to the doctor. It turned out that fly larvae had gotten into his hair and made their way under his skin. They grew and became worms! The doctor said it was nothing. He told us to apply Vaseline, that then the worms couldn't breathe, and when they came to the surface we should just squeeze them out. We did it, but quite a repulsive introduction into African life and in some ways indicative of the many challenges we would face.

Our shipment from Spain had arrived and Eric began to push Customs for a reason that it had not been released, and we expected they wanted *dash*. However, they just kept putting him off until he was finally informed that he was suspected of smuggling jewels! We were dumbfounded, guessed they were talking about some of his mineral collection and told them they could keep them, but give us our household furniture and possessions. When they were released and Eric asked what 'jewels' caused the delay, the Customs official was obviously embarrassed. I had bought some sparkling Czechoslovakian crystal beads while in New York, the necklace had broken and when I restrung them, I put the leftover beads in a ring box. Admittedly they were bright and shiny and at a glance they might look like diamonds, but with a hole right through the middle? Now we could leave the hotel and move into our new 'home'. Life was good.

31. Eric: Ikoyi Club and Okan I!

One of the first things we decided to do was join the Ikoyi Club, a British Club that was open to all expatriates and Nigerians but used mostly by expatriates. The Club had recently been transferred to Nigerian management. They had an 18 hole golf course, a swimming pool, tennis courts and was a favorite stop for a beer or gin-tonic after work. At night there were British movies shown outside, weather permitting, occasional dances, plus other social activities. They had a wide-ranging library that emphasized British literature, the kind we often see now on PBS Masterpiece Theater. The boys were introduced to Gerald Durrell's nature books, John learned to play golf and began judo, all the boys swam and Libby played tennis. I struggled with golf.

The British were phasing out their imperial presence in Africa and regarded us Americans as rather uncouth and overpaid competitors. One of the first obligations we had to learn was to drive on the 'wrong' side of the road. All of the vehicles, including company vehicles, had the steering wheel on the right-hand side, which often caused us embarrassment and hair-raising escapes. There was also a dress-code that most of the English civil-servants observed: shirt, short pants, and knee-high stockings all had to be white and women always wore skirts except to play tennis or swim. There was another Social club, limited mostly to executive types and government officials, to which we were occasionally invited. There was also a sailing club and a motorboat club and there were no color-lines in any of these clubs.

In the office we were trying to imagine the geology of six thousand square miles with about a dozen electric logs from the wells that had been drilled by mostly Shell BP. Electric logs are profiles of the rock formations that were penetrated by the drill holes. The British

Colonial Administration had limited access to the area for geological exploration and information was very sparse. Essentially, there was little subsurface information available, neither Gene nor I had worked in offshore drilling, and it was obvious that we would be completely dependent on Gulf's geophysical data. After the Blocks of one thousand square miles each were awarded, Gulf did a geophysical survey in 1961 and 1962 and this would essentially be all the information we would have for our forthcoming drilling program.

Geophysical crew recording in the delta.

The geophysical surveys were carried out by GSI, Geophysical Service International, under the direction of Gulf Research and Development Co., (GR&DC Harmarville, Pa.) which included Jake and Darryl. Their interpretations suggested a number of structures that appeared to be good drilling prospects. Gene and I picked several and were now waiting for the Caltex rig to be available. Garland and the drilling department had decided that the closest facility that we could use as a shore-base was at Burutu, an African palm-oil and rubber shipping port and trading facility on the Forcados River. Gene and I flew to Warri and then took a riverboat, which carried freight and supplies, and was the only method of transportation to Burutu. I was anticipating seeing something like the Disneyland tropical world, but saw nor heard any river-life; no monkeys, birds, crocodiles or hippopotami, such a disappointment for me, but it had been good eating for the locals.

Garland had somehow managed to get everything together in this strange world with the coastal freighters and through the heliport. We spent two nights at the Burutu Base in the guest quarters of the Port where the administrators lived in bachelor quarters. It appeared bridge and billiard tables were their only recreation. This was a United Africa Company operation, well-run and we felt conspicuous in our travel khakis. They dressed for dinner in fresh white shirts, shorts and knee length white stockings just like the Ikoyi Club in Lagos. These formalities helped the Colonials maintain their empire and the respect of the local people but was not our style.

In November the rig was finally released to us. Caltex had drilled the first of two locations that they had selected, but opted to discontinue their drilling operations. Under the terms of the Nigerian drilling regulations, they could suspend the final completion of their hole and make further evaluations of the geological results of their test. The two Caltex Blocks were at the very nose of the Delta, about one-hundred and twenty-five miles from Nigerian Gulf's prospecting licenses. The rig was a 'jack-up', meaning it had to be towed by tugs to the position we had located in Block D for our first test, taking about three weeks. In the meantime we had to select future locations since our obligations to the Nigerian Government required a minimal amount of geophysical and drilling operations.

My job was to locate a buoy seven miles out in the ocean at the selected position! We decided that since there were no references onshore, we must have a survey to tie the shoreline to the geophysical map. We hired Herb Hamm, a Swiss surveyor who determined he would need several reference points (monuments) onshore and we hired boats and personnel to assist him. He somehow managed to locate a buoy in what he considered to be our selected position. We did not have the Global Positioning System that we have today and I realize now how fortunate we were. I recently spoke with George Livo, a retired Gulf geophysicist in Denver, who told me that our monuments had been destroyed by locals almost immediately after we used them. The local fishermen were suspicious that foreigners were trying to take their land.

We positioned the rig in thirty-five feet of water, the shallow depth at this distance from shore was indicative of the continued flat and featureless onshore topography. The rig had several levels that contained crew-quarters and galley and the drillers were all from the States working for The Offshore Company. We understood that the base cost was twenty-five thousand dollars a day, typical for an overseas offshore operation in 1963. We began drilling in early December and John and Whit would be doing the well-sitting, and offshore well-sitting was new to them also. We had a small well-logging unit on board to monitor the cuttings and a permanent Schlumburger electric-logging unit with engineer. We also had several auxiliary boats that could carry equipment and personnel as well as a heliport.

The operation was twenty-four-seven and after the surface casing was set, John and Whit alternated a week at a time. By Christmas we had drilled about six-thousand feet and had begun to see oil and gas in the cuttings. We decided to run our first electric-log surveys and they confirmed that several zones looked productive; we continued down to nine thousand, ran our final electric-logs and set casing to the total depth. By New Years we were ready for the final testing and I flew down from Lagos to be aboard after the cementing was finished.

There was high anticipation and anxiety. We knew that executives from London and Pittsburgh were just as anxious and we were, probably more so. We hoped this would be another oil-boom that would help the industry diversify the overseas oil production. It was becoming clear that the Middle East would be the center of world oil production. After the aborted 1956 invasion of Egypt by Britain, France and Israel, responding to President Nasser's nationalization of the Suez Canal, it was obvious that world oil resources would be threatened by geo-political pressures. President Eisenhower stopped the invasion by threatening to cut off oil supplies to Britain and France, understanding that this would create a cold-war threat from Russia and Nasser agreed that he would not close the canal for political reasons.

My job was simple on the rig at this time; to select the best sands to be perforated and be tested, to determine the oil productivity. I looked at the Schlumberger logs, picked the perforations, talked with Gene on the radio and he agreed. We perforated and Garland rigged the test tool. The tool was opened and we got salt-water. Back to the drawing board and in consultations with Lagos, we decided to try the next best looking sequence. Garland had to cement the previous perforations and then we tested the next best sand, somewhat lower, and again got salt-water.

I was getting desperate and Garland was grim. I was determined to try the upper sands again and we decided there should be enough of a seal after several cement jobs. We perforated, Garland began the test and I went back to the galley for lunch, feeling confused and depressed. Eventually Garland found me and said "We've got oil!". We called Lagos and they wanted to get the best possible test without damaging the hole. Garland was able to establish a flow of 2,300 barrels a day. Okan 1 it was. Hallelujah and Happy New Year!

32. Libby: Mini Minor, Schools, and JFK

We moved into the house allocated to us on Ikoyi Island, the inner of two long barrier islands parallel to mainland Nigeria, and the first thing we noticed was the blessed reduction of noise level and occasional breezes. The doctors returned and each had reached the same diagnosis for me: brucellosis, undulant Fever, Mediterranean or Malta Fever, all the same devastating illness that sometimes cannot be cured. Apparently I had picked it up eating that contaminated whipped cream on strawberries, from diseased cows, that happy night in a bar with friends in northern Basque Spain after hearing Segovia play his magical guitar. A very sad result from such a wonderful evening. The New York doctors hadn't even attempted to grow cultures for such; the thought being that since it was essentially eradicated in the USA, that it had been eradicated worldwide!

Both doctors telegrammed England for the latest treatment for brucellosis and they agreed on a powerful combination of drugs. There was a huge quantity of enormous sulfa pills that I had to take along with some very strong antibiotic pills each day, and I was to receive a streptomycin shot daily. The doctor who gave me the shots, the cantankerous one, had offices further out on the island where I had to go each day for several weeks. At this time I was struggling with new servants, new acquaintances, new surroundings, complaining wives whose husbands worked for mine, a completely new foreign culture, unfamiliar foods with horrific shopping conditions, not to mention new friends for the boys and all the ramifications of different schooling, all in a climate often described in British literature as 'white man's grave'. Plus a husband totally absorbed in the possibilities of oil discovery to which I was taking a back seat.

Two weeks into this treatment, and under these stressful conditions, I was a nervous wreck from the medications. During that month in the hotel, Eric had bought a red Mini-Minor from a departing Company friend and at that time the Company allowed wives to drive in Nigeria. I was returning from my shot at the doctor's office way down Owolowo Road when I was startled with sirens and a black police Land Rover stopped me. Two enormous Nigerian policemen jumped out and appeared on either side of my tiny Mini-Minor. "What you doing, Madame? Your flicker is flicking and has been flicking for long time!" I probably would have had to hide my mirth in normal times but this final insult to me was just too much and I broke down crying.

I sobbed that I was sick and had to go to the doctor every day for a shot. Well, if they could have gotten in that little car and held me, they would have, but instead they said: "Oh Madame, we so, so sorry! We will take care of you! Follow us!" And so, with me crying, and their sirens blaring, they accompanied me home. I don't remember even smiling once at the absurdity of it all. This was only one of many times that I found myself the recipient of Nigerian gentle understanding and compassion. Over and over, they proved themselves to be a happy and warm people despite their hard lives and often primitive living conditions.

There were two bedrooms and bath for the boys upstairs along with our master bedroom and bath, and miracle of miracles, all three bedrooms had air-conditioners that worked for the three years we were there. There were ceiling fans downstairs in the living-dining room and the air-conditioning would drop down from upstairs if we kept the heavy window and door curtains closed downstairs. However, that produced a stuffiness that no one really liked. There was a long kitchen, the depth of the house, with an adjoining bathroom and then an outdoor front porch that ran the length of the house with iron grille and gate. Outside there were some banana trees but this was new construction and we were on a corner with a row of houses between us and Owolowo Road and across Ademola Street was another row between us and Five Cowrie Creek. We were sent a gardener, David, and I asked him to plant grass and a row of evergreen trees around the corner to give us privacy. However, despite my protestations, he insisted upon planting sticks in the ground. To our utter amazement, in six months we had 3 and 4 foot high trees growing and hybiscus bushes just seemed to appear overnight. The weather was always humid, the temperature hovered between 75 to 95 degrees and you always knew that 75 degree winter weather had arrived when the Nigerians began to wear wool hats.

I began with one Ibo servant named Bright. He was to be my cook, a misnomer if there ever was one, but Bright was honest, dependable, intelligent, and somewhat educated. He had a wife and children who lived somewhere on the mainland and whom I never met in three years. He got on his bicycle when it was still dark in the morning to make his trek from the mainland to our home, and returned after he had cleaned up our kitchen at night. We had servants quarters and although I wanted him to live there, he would not. Bright was what I would call 'an arrogant man'. He tolerated me only because I was his Madame and we paid him. It was soon apparent that we needed more help and he brought us dear Matthew. Bright would bring food from the market on his bicycle, and peddlers came from door to door, but shopping was a problem, and with no telephones and little boys always running around the neighborhood, we needed more help.

Matthew was a bright young man of eighteen who would do anything for our family, and he watched over the boys like a mother-hen. Like Bright he was an Ibo and these were Christian folk. Women servants were only available as 'nannies' in Nigeria. We also had servants outside but they were mostly Yorubas of the Muslim religion. We had to have an outdoor *watch* (watchman) at night whose name I never knew. He had his prayer rug and sat or slept outside our house (as long as he thought we were awake) to protect our house from *de teevs* (thieves). Actually, he was just part of the system of *dash*. The

Company's general contractor, Sarumi, produced him and paid the thieves to stay away.

Our gardener, David, was a faithful and gentle man, and my driver, Fred, was always there when needed. The Company people were told that after an accident that involved Americans, the Company would not allow us to drive for a while. We were also advised that if we were ever in an accident we must keep on driving and not stop. In this accident that had resulted in a death or serious injury, the locals had killed the driver on the spot. Then the person who they thought had been killed, began to stir and the crowd was so mad that they killed him also. And so, I had to have a driver in my little red Mini-Minor. I struggled to get into the back seat the first time we drove out, which was expected of me, but that uncomfortable, sweaty trip was quite enough. Fred was most embarrassed when I decided to sit up front and as we drove through the city, the Nigerians laughingly shouted at him, about that blonde woman sitting with him, and I'm sure he turned red under his very black skin.

I look back on these Bolivian, Spanish and Nigerian servants that helped me and my family survive in these foreign countries and I am humbled. I still remember each of them from so long ago as friends, many as good friends and some I loved, people without whom I could never have survived. We Americans were embraced in every country and a sadness I must live with is that few, despite their intelligence, could write or afford the time or pittance for lessons to learn to write. The scarce letters we received were generally written by others with the embarrassed acknowledgment that they couldn't write. However, our Matthew studied every night in our kitchen trying to learn to read and write. I wish I had been able to find more time for him.

The boys were soon students at a grammar school called Corona. It was run by the English in the community and at least one third of the students were Nigerian. We decided we should only speak Spanish between us one day a week - at dinner on Friday nights - so the boys could wean themselves from Spanish expressions and begin to sound like English folk. I had managed to get Paul ready in English for the third grade with the Calvert system in a couple of months, John's English was okay, and I continued with the equivalent of Kindergarten with Mark. John was soon taking judo and playing baseball, and Paul was chasing bugs and lizards with his new friends everywhere. He soon became friends with a Nigerian boy in school who turned out to be the son of the Prime Minister of Nigeria. We didn't realize he was Sir Al Haji Abubakar's son until Paul received a formal invitation to his birthday party. The day after the party there was a picture on the front page of the newspaper with Paul and his friend Jan, the son of the Danish Ambassador, carrying the birthday boy. We still have the picture.

One day, I crowded four little neighborhood boys along with Paul and Mark in the back seat of the Mini-Minor while Fred and I sat in front. The boys wanted to go to the enormous Jankara market on the mainland to look for chameleons. On the drive, some of them began talking about monkey's paws and other strange 'parts'. Fred parked the car and promised me that he would lead us through these acres of produce, raw meat and every kind of available item for sale in Africa. Incidentally, meat had to have its skin attached because it was not uncommon for human meat to be sold. These factoids excited little boys but made all we adults very uncomfortable, and I can't recall ever returning to Jankara market again. The boys found everything they wanted and more, and I know they had a dried monkey's paw tucked away for years.

One of the memories that remains with me was what looked like a small city block covered with drying cockroaches. Fred informed me that they were good medicine. They were pulverized into a powder and applied to an infected area. I read in Time magazine several years later that, indeed, in a Yale study, (I think), cockroaches carried some antibacterial powers. There was so much to learn in mysterious Africa and with our oil discovery, it was obvious we would be around for a while.

The Ademola Street kids.

Neighbors who immediately became a part of our lives were Jake and Harriet, next door, and Jim and Doris Buchanan, African Americans who lived across the street. Jake and Harriet had two little girls and Jim and Doris had three boys and one girl, all within the years of our boys. Gene and Alice lived a bit further down the street and their daughter, Sharon, was John's good friend. Other good friends of Paul's, Jerry Schneider and Johnathan Welles, lived on the street. The daughters of an Ethiopian official who also lived on Ademola taught our boys a new dance called the 'twist'. It was a lively neighborhood.

One night we were playing bridge with Harriet and Jake, and John came running downstairs from the TV in our bedroom where he was allowed to watch *Rawhide*, shouting that President Kennedy had been shot. We were stunned and really couldn't believe him.

We watched as the whole terrible scene unfolded and when he was pronounced dead, realized we had to inform the neighbors since no one had telephones, and few had televisions since there really wasn't much to see. We decided Eric and Jake would go down the street and Harriet and I would go in the opposite direction. It fell to Harriet and me to go across the street to Jim and Doris, although we felt that Jim must already have been told, as a Colonel and US Military Attaché with the US State Department. We knocked, Jim came to the door barefooted, and we heard some great jazz playing inside. We told him what we had seen and I shall never forget his comment, "My God, pray it was an American!!". He began shouting and pulling on his shoes at the same time. In minutes he had left the garage with his chauffeur, who was always standing by. As we made our way up the street, the news began to spread. I received a letter from Ossie's wife in Spain a week later, writing how shocked they all were, and asking for our opinions of Lyndon Johnson. The Europeans had been so proud of our last choice, President Kennedy.

33. Eric: Okan Field, Sarumi, and Snakes

After the euphoria of our first discovery, we had to begin to develop and determine the extent of a possible new oil field. We selected a second location, a mile further south on the seismic structure and moved the rig to drill a confirmation of what would be called Okan Field. In the meantime, we had picked a location southeast of Port Harcourt in the swampy area to drill a wildcat (exploration test). We had seismic data that indicated a rather small structure near offshore and we would be able to drill this with a swamp barge, a company owned rig, that would be brought from Louisiana.

Garland, John, Whit and I took advantage of this waiting period to travel to the edge of this swampy area in eastern Nigeria to scout the surrounding ground area. We visited a village of small wooden huts with either tin or palm-leaf roofs; the people were friendly and as curious of us as we were of them. Shell BP had done some drilling and exploration work in the vicinity and although they found no production, the local natives had been able to find work. We left the countryside with a generally optimistic feeling that we could find good labor if needed. As it turned out our first test in this area, Tubu 1, was located just offshore and was successful.

It was obvious from the very beginning that we would be hard-pressed to meet all of our obligations to the Nigerian Government. We were also receiving a great deal of pressure from Pittsburgh to export oil as soon as possible. They wanted us to produce 25,000 barrels a day before we would start exporting oil. Our agreement with the government obligated us to develop a Nigerian staff so that at least ninety-percent of the Company would be Nigerian within ten years. That meant we had to find suitable Nigerian managerial people as well as field laborers pronto. Besides local requirements, we needed experienced oil people and fortunately, our large operation in Venezuela, Mene Grande, was in a mature phase and had well-trained American and European geologists and engineers available to us.

I had never worked with Africans before, but I felt comfortable with them. They were happy, courteous and enthusiastic, however, often overestimating their abilities. From the beginning of this Nigerian adventure, I wondered how the southern-born exploration and drilling people would react when they lived in a new country that was run by Africans. My apprehension was relieved to some extent shortly after my arrival when Jake, Darryl and I received an invitation to attend a post-Ramadan feast by Sarumi.

Sarumi was a contractor who provided various laboring type services in the Company. He furnished us with our gardeners, night watchman, stewards, drivers, and found people for all general services. We were picked up at the hotel, brought to his home and we made small-talk all afternoon while a goat was being cooked. We had been invited to witness the ritual slaughter of the goat, which was accomplished by slitting its throat. Jake and Darryl, who were not from the deep south, but from Missouri and Oklahoma, had taken the ritual in their stride and showed no problem with sitting down to dinner with a black Yoruba Muslim and several of his male Nigerian guests.

It was curious to me that although we had met several Company employees who were Christian Ibos from eastern Nigeria where there was oil production, our first invitation into a private home was Muslim. The British had been careful to create a governmental framework that would accommodate both Christian and Muslim populations. The President of Nigeria was a Christian Ibo from the east, the Prime Minister was a Muslim from the north, while Lagos and the west were predominantly Yoruba Muslims. There were more than one-hundred other tribal and linguistic groups but English was generally understood. There was also a strong thread of Pagan religion throughout the country that was practiced along with Christianity and Muslim faiths. The oil production that we expected was regarded as being in the western region, and this generated much excitement in Lagos, which was primarily Yoruba.

If we found the production that we hoped in this first field, this would be a significant find for Gulf, who needed to diversify its source of oil. Our main sources of production outside of the United States were in Venezuela in South America and Kuwait in the Middle East. Kuwait, that we partnered with British Petroleum, was the largest field in the world until surpassed by Saudi Arabia. Because of the increasing hostility between Israel and the Arab countries, fears were beginning to develop that oil production in the Middle East could be interrupted. We were also concerned with the Soviet Union's influence in the region and their ties to Egypt and Iraq. Libby and I were invited to several Embassy parties that often included members of the Soviet Embassy and our neighbor, Jim, told us that at least several of these people were KGB. The Soviet Union was entering into a trade war with the US at this time, competing to sell heavy equipment in the developing world and to spread their political system.

The confirmation well, Okan 2, was successful and we were off to the races. The Nigerian Ministry of Mines and Power requested a weekly report from the Company of progress in the field and Gene asked me to take the reports in person. I gave my reports to a man named M.O. Feyede, the first assistant to the Minister, a Nigerian from the west, who had been educated in Britain as a petroleum engineer. M.O. had worked for Shell BP for five years in the Delta, and was familiar with oil and gas operations. He was friendly and knowledgeable and it was obvious that the government was extremely interested in our production. When Nigeria's oil production grew large enough to be qualified for membership in OPEC, M.O. became Nigeria's first representative that generally met in Vienna. Eventually he became the Secretary-general of OPEC and served two terms with distinction. I later learned that he was a hereditary Chief of his tribe.

We had no servants on Sundays and early one Sunday morning while Libby and I were still in bed, Mark came running up the stairs, shouting that John and Paul had cornered a cobra by a bush near the water between the two houses across the street. We both moved fast. We met them coming across the street, crestfallen that it had gotten away. It was obvious then that we had to teach the boys about just how close they could have come to a sudden end. Africa was filled with poisonous snakes, some of the most poisonous snakes in the world and Paul, especially, was crazy about all reptiles.

I had a connection with the Nigerian Field Society who had asked me to speak to them about geology and our offshore oil find. One of them was an American Missionary, a Reverend who had a large collection of poisonous snakes whose venom he sold to laboratories in the States to support his mission. I told him of our concerns and he invited us to his home on the mainland, to meet his wife and three little girls, and see his snakes. We included Jim Bascom, the Company doctor, and wife Jeannie, on our expedition with the boys, which was about an hours drive to the edge of the jungle. Libby remembers the huge bougainvillea that encompassed the house and the refrigerator on the porch that was black with mildew on the outside. My first impression was of Oola the Chimp, who was chained on the front porch. The boys were thrilled and Mark, being the youngest and least cautious, immediately ran to the friendly Oola who grabbed him and gave him quite a scare.

The Reverend showed us his herpetology laboratory and assured us that he had anti-venom serum in the refrigerator. He had an assistant who helped him with his snakes, and there were large lizards, in sunken cement tubs, some five-feet long. He also had several large pythons outside, six or seven-feet in length, inclosed in a screened area. The poisonous snakes were in several stacks of wooden boxes with glass doors on the front so they were easy to see. He explained the extreme danger of some of these snakes and in particular the spitting cobra. At one time there had been a crack in a glass and his assistant looked into the glass, the cobra spit through the crack, and he was temporarily blinded. They would blind their prey to immobilize them before the kill. There were several varieties of cobras, black and green mambas, and other kinds of poisonous snakes. The Reverend explained that he never lacked for snakes for when the locals found them in their fields or homes, they would send a runner for him.

Oola likes Mark.

Finally, he showed us his vipers, the worst being the Gaboon viper, with a dappled skin of rust and grey coloring which allowed them to blend into the forest floor. He took a seven foot long snake hook and opened the closed door of a wooden box on the ground.

Inside were several vipers, just a dozen feet from us. These vipers had fat bodies, with a tapered tail and with a large, diamond shaped head. He claimed they were very lethargic and slow to move and we were mesmerized and terrified at the same time. Jim B turned white. The boys got the message. At a later meeting of the Nature Club, I was invited to join them on their next field trip which would be at night, in a swamp, and with some kind of made-up excuse, I declined.

Reverend Bauman the herpetologist.

34. Libby: Shopping, Femmi, and School

It was very hot on our first Christmas in Lagos. We had a choir group at the Embassy, with whom I sang, and the Nigerians turned some of our traditional songs into wonderful entertainment with their drums. The celebrations were beyond anything we could have imagined. Male Nigerians roamed the streets in dance groups, some wearing feathers and head masks and others with painted bodies and faces, carrying spears, and beating drums. About twenty came to our house, were laughing, shouting, and singing in different languages and then finally killed a chicken. They spilled its blood around and then after a serious moment, there was a great deal of laughter. We believed these were disguised Yorubas from the Company but have no idea what it was all about. Pagan beliefs were all mixed up with Islam and Christianity. I suppose they felt doubly protected with both.

Jim and Doris had access to the State Department commissary and gave us a beautiful frozen turkey from the States that Christmas, for which the whole family was MOST grateful. We had to buy our expensive meat at two international stores in town, or go to the open market, but it was a roll of the dice there as to what you were buying; same with the vendors who would come by your house. Frankly, I only felt good with live chickens. We soon learned why the British were so fond of curry and that was one thing that Bright had learned to cook well. He had difficulty eating our American or European cooking, Matthew did better, but mostly they wanted their *garri*, ground from a white root, our chicken, and sometimes vegetables.

After Christmas, I answered the door to find a tall, young Nigerian boy who asked me with a very British accent "I say, Mrs Ericson, could the boys come down for a cup of tea?" He lived down the street, was home from school in Britain for the holidays, and Femmi Pratt became a friend of John's. His mother was the Director of Nursing and Public Health for the Administration and his father, a doctor, was also involved with the Administration in Public Health. The boys always looked forward to his vacations home. Some years later when we were returning home after five years in Sydney, Australia, our last foreign assignment with the Company, we had a five-day stop in Moscow with Paul and Mark, now 18 and 15. It was 1974, the cold-war was at its height, but friends who had lived in other foreign countries as we had, said it was a 'must' stop on our way home.

Christmas carolers on Ademola Street.

The Russians did everything to make our life uncomfortable, spoke only Russian and were pointedly rude. Danish journalists at the hotel were helpful with English at times, and once we even lucked into some shoe-salesmen from Spain. One morning I saw two tall Africans shopping in a store who were speaking Russian but looked so Nigerian that I approached them and asked if they spoke English and could they translate for me. It turned out that they were Nigerians studying engineering and medicine on Soviet scholarships and had a three day visa from their universities to meet each other in Moscow. They were friends of Femmi and told us he was beginning to study psychiatry in London. We took them to lunch and they bought champagne. They were as delighted to meet us as we them.

In Lagos, to get into UAC (United Africa Company general store) or Leventis, a Lebanese department store, one had to step over and around dozens of cripples. They all had their hands out begging. Most were clean and some well dressed, but they were a terrible sight to behold. To make beggars of them, their bones had been broken in childhood by their impoverished families, and their twisted arms and legs made it impossible for them to walk. They crawled around like--and I hate to even write it--like crabs. At first I didn't know how to handle it. Some people completely ignored them, said they were rich and had no business begging, some stared painfully straight ahead, others just threw some money. Most of them spoke broken English and how one could close your mind to their pleading was beyond me. I finally decided to carry a hand full of coins every shopping day and distribute them until they were gone. I imagine good positioning along the sidewalk had its price.

One afternoon I was caught in one of those sudden drenching downpours and waited inside Leventis until the worst passed. The cement sidewalks were two-feet high and the water rushed through the streets almost up to the sidewalk. I was hurrying along when I saw an Oba (Chief) with his entourage approaching me, holding umbrellas over the Oba. They were laughing and talking, totally ignored me, and guess who had to get down in the water until they passed? I'm not sure exactly what they thought of us, but I know they considered our

African Americans to be sons of slaves and therefore inferior, and of course, I was a woman.

One day I heard the screeching of young boys in our garage. I ran out and found two young boys about the age of John, hopping around on Eric's work table, while a man beat their legs with a stick. I demanded to know who, what and why in OUR garage and was told under no uncertain terms that they were HIS slaves and they were misbehaving. I knew he was an employee of Sarumi and I was so indignant that I insisted through the Company manager that he be fired. I doubt that he was fired but he never came my way again. Most of these young children were sold by starving families who lived in the jungle villages.

The Company decided that our cooks must have a lung x-ray for tuberculosis and I informed Bright that he must go to the hospital. It wasn't far away but he was gone a long time. When he finally returned, he simply said they would not take his x- ray. I assumed he didn't know the difference between 'would' and 'could', that they were busy, and I sent him again in a few days. He returned with the same negative results and responded to my query with, "Madame, there are bad people in the hospital". He lifted his nose and straightened his spine even more than usual. I kept pressuring him why and he finally made it clear that Muslim Yorubas ran the hospital and he was a Christian Ibo and they did not like him. So I had to take him to the hospital and wait for him to be x-rayed. It seemed they had just received some new x-ray machines and while in the waiting room, a technician (?) began aiming a machine here and there and laughing with another man in their language. I quickly stepped outside.

I turned the downstairs bathroom, off the kitchen, into a nursing station and dispensary. There was always some infected cut that needed care by one of the servants employed by us, or a gardener in the neighborhood. David came to me one day with a fever and nasty infected and swollen abscess on his leg, and when I pressured him for a reason, he told me that he'd been given a shot by the local medicine man in his village to increase his sexual prowess! I could do nothing with something that serious and took him immediately to the hospital. One week later he was returned in a big open truck along with a load of other patients, who were all dressed in what looked like black and white stateside prison uniforms. Quite a sight. Fortunately, he had survived and was very pleased that he had.

The British often sent their children to England for schooling anytime after kindergarten and there was no English schooling after the sixth-grade in Lagos. The Americans had been working at organizing a school for about a year. They brought a Principal and some teachers from the States and The International School opened about six months after we arrived. The rest of the teachers were hired from the many expatriates in Lagos but the scheduled second-grade teacher, whose husband had been transferred, had to leave almost immediately after the school opened. I had offered to help in any way when the school was first begun, and was called as a substitute second-grade teacher until they found a permanent one. I accepted since Mark was in the first-grade by then but shortly the Principal told me that they wanted to move both Mark and John up a grade.

Eric and I discussed this possibility but John was always a very hyper child from birth and we really didn't want him to be advanced and under more pressure, and I did not want to be Mark's teacher in his first school experience. However, the Principal insisted. The family discussed all the possibilities at dinner and the boys decided that if the second-grade students didn't know that I was Mark's mother, it wouldn't put him on the spot, and John, although young for the class, was very eager to move into the sixth-grade. By this time we had settled into our house, had a routine with the servants managing many duties, and so I said, 'yes' to both advancements, thinking I would be teaching Mark's class for only a few days.

Hiring a permanent second-grade teacher became difficult but I was really enjoying teaching these children. The precious little girls would cling to me and I was totally at home with boys, so all went well for several months. Then one day I took them outside for our mid-morning break to read a chapter of 'Alice in Wonderland'. There was a wall being constructed nearby about three feet high and the boys would climb up and jump off saying, 'Look Mrs Ericson!' Then Mark got up, jumped down, and said, 'Look Mom!' There were about fifteen children in the class and they were all silent and looked serious, but then began to ask "Is Mark really your little boy?" over and over. So I explained that Mark was afraid he would be considered a 'teacher's pet' if they knew. Two little girls had tears in their eyes. I had a lot of making up to do but within a day or so they had forgotten that Mark was my son. He was well-liked which helped, and fortunately the second-grade teacher showed up in another month.

I continued to substitute but only for a few days at a time. Math in the upper grades was tough for me and the students knew it. One Christmas I asked the seventh-graders to write on the blackboard every different holiday greeting they knew and they wrote down greetings in at least fifteen different languages and there were many religions represented. The English had established themselves further out on the island but in our new neighborhood alone, we had Australians, Finns, Danes, Ethiopians, Canadians, Americans, including African Americans, as well as Nigerians, among others. The students were bright and all spoke English with little accent but came with many different features and in many different shades.

When the American expatriate community began to grow, Gulf, with the biggest stake, was a leader in many groups, such as the Embassy, Peace Corps, United Nations, and private NGO's. Eric was asked to represent Gulf on the board of the newly organized school and this broadened our contacts within the community, social and otherwise. At one time, we were invited to a large formal affair at the home of the Prime Minister Abubakar for Julius Nyere, the Prime-Minister from Tanzania, a larger than life leader in East Africa. Both Abubakar and Nyere were dignified, intelligent and outstanding men at a time when political violence was probably at a minimum in Africa.

Certainly not comparable to the Africa of today. We were also invited to celebrations in villages, with dancing and feasting, sometime the only whites, and treated as honored guests. Many times we suspected that our association with Jim, our neighbor, was responsible, but Company Nigerians saw that Eric was not racist and was genuinely trying to help the Nigerian geologists and others to learn the business.

One day I saw police cars driving into and parking around Jim's house, but there was really no opportunity to find out what was happening. Harriet and I wound up on her balcony, watching, and the men joined us when they came home. Through the children and servants we had found out that it had to do with Jim's cook and somehow murder was involved. Jim and Doris later told us the grisly story. Their cook discovered that his wife was having an affair with a policeman, he consulted a medicine man, and they decided to kill him. The policeman agreed to meet them in a mangrove swamp and after their dirty deed, carved off his toes, fingers and the other protruding parts of his body and brought them back to be sold in leather pouches because he had special powers as a policeman! The pouches could cost as much as 50 British pounds. In the meantime, they kept the parts buried underneath the outdoor fireplace in the servants quarters. Either the medicine-man or their cook tried to sell the policeman's bicycle and got caught.

This cook, a very polite and pleasant man, a 'good Catholic' who went to church with his wife and children, often came to our house looking for Doris's boys. Our boys had often eaten his good cooking at their house and we were all horrified and yet titillated with this bazaar story. The cook and medicine man served a little time in jail but were out before we left Nigeria.

35. Eric: Oil production, Ju Ju, and Sarumi's party

With the completion of Okan 2, we knew we had a field of considerable size and Gene and I were trying to figure out where to go next. We had more manpower on the way and found a larger suite of air-conditioned offices at Tinubu Square, more or less in the center of the city. A reasonable estimate of the size of the city would be 'less than a million people'. The center of the city and square was continuously loud from early morning until late at night with voices and music. They were a happy people and had reason to be since they were enjoying their liberation from Colonial status. I was told when we arrived that Nigeria grew enough food to feed its people but soon the rural farmers moved to Lagos looking for new opportunities. Foreign companies, European and American, were flocking to this newly independent African nation, the largest south of the Sahara, but there just weren't enough jobs to go around. The population then was about forty million but it exploded to one hundred and twenty-nine million in 2006, 50% Muslim, 40% Christian and 10% indigenous beliefs.

Glenn, the general manager, was back in the States much of the time dealing with all the new Company business. Gene was in charge of the exploration and development geology and Charlie was running the office, staff and business affairs. We finally located one Nigerian geologist but geophysical graduates were virtually nonexistent. We had to bring in more drilling rigs, a production manager, personnel manager, a materials man, supply personnel, accounting and more senior production and drilling people. It was also obvious that we needed a permanent shore base which we located on the Escravos River, much closer to our Okan field.

They also sent an export facilities advisor who would oversee the building of a temporary export terminal that would be in water deep enough to load large tankers. This floating terminal would consist of two large tanker hulls that had been welded together in Japan, into one giant storage vessel. I marveled at the explosion of activity and was dumbstruck with the ability of the Company to move so much so fast. I later found out that the CEO of Gulf Oil Corporation was soon to retire and this successful new venture in Africa would certainly enhance his retirement.

The swamp-barge arrived and the drilling began southeast of Port Harcourt on the coast, Tubu I, was successful although not as promising as Okan. We now had two fields two-hundred miles apart and we would need more well-site geologists as the fields were developed. Mike Kontz and Barry Shelkin came down from Libya where exploration work was winding down, Mike Brady from Mene Grande, and P V Ferguson from New York. Eventually we had about ten geologists in all. Lemuel Ezogu was our first Nigerian geologist but after a year we decided to give him a scholarship and send him to San Diego State for more study. Before I left in 1966 we had three Nigerian geologists.

By the end of 1964, we had drilled about 14 producing wells in Okan Field, had located two more offshore fields, Delta, between Okan and the shore line, and Meren, located about fifteen miles northwest of Okan. Meren looked even larger than Okan. The wells in Okan Field had established the production potential of 30,000 barrels of 35 gravity (sweet) oil a day, for which there was a growing market. We were now waiting for the export-terminal to be completed where large tankers could be loaded, as well as for the pipeline from the field to the export terminal. Concurrently, work on the Escravos terminal would be finished in a few months.

Now we had a new general manager, Newby Simpson, since Glenn had to remain in the States with Ann, his ailing wife. Glenn had left an important legacy, had impressed the Nigerian officials that he could be trusted and that the Company operations would be carried out quickly and efficiently. Newby was a petroleum engineer and former vice-president of production for Gulf domestic, whose mission now was to get the Nigerian crude moving to market. He had to deal with the Nigerian Government that had had a long standing relationship with Shell BP, and they seemed to be moving slowly with our new company and new manager. This was very frustrating for all of us since we were ready to ship oil and were now being delayed by bureaucracy. Different parts of the government wanted a piece of the pie. Northern and western Nigeria wanted a share of the income with the western Delta tribal area. Despite all the interests to turn the valves and open the flow, we were not able to begin exporting until the first week of April 1965. Considering that we had completed the first discovery in January of 1964, this was a world record.

Nigeria for its part would receive fifty percent of the oil production, with additional taxes on the operation, and new jobs were being created for the people. Gulf's obligations to the country, in addition, was to spend ten million dollars in exploration and those expenditures had mostly been met. At this time, the market was flush with oil from the Middle East, a barrel of oil was about two dollars and the United States was the largest oil producer and consumer in the world and able to control the price. After oil production in the USA peaked in early 1970's and began to decline, OPEC (Organization of Petroleum Exporting Countries) has been able to regulate production and control the price. A few months later in the New York office the Vice President of Accounting asked me why were we drilling for oil when we couldn't sell what we had now!

The social obligations grew along with our success. There was an increasing number of cocktail, dinner and beach parties, to greet visiting Gulf dignitaries as well as invitations to Embassy affairs and to welcome new oil people. One large dinner party was hosted by Glenn before he left, at the new Chinese restaurant in Lagos that could feed about one-hundred and fifty people. The Chinese family who owned the restaurant told us that it had taken one month to train their waiters. At Glenn's farewell party, there were probably seventy-five Company people and other guests. David Garrick, the Nigerian director for Gulf Oil, an attorney, had been educated in a British Prep School and at Cambridge. He was a very handsome man, was a member of the royal house of Benin, and was there with his Irish wife. Libby didn't see him at first, and later when she did, said, "Oh Hello, David, I didn't see you before." His response was memorable. "No wonder, Libby, I was standing in a dark corner." He was a witty man.

Once when I was on one of the offshore rigs, David G visited the rig and went to the mess hall for coffee. A recently arrived driller from the deep-south, walked into the galley, saw David sitting there and said loudly that he would not eat with a N----, turned around and left. David was comfortable anywhere in the world and whether he complained or others did, this driller was immediately taken off by helicopter and on the next flight out to the States. The Nigerian people were a happy, confident and proud people and had no sense of inferiority or animosity for other people, only with other tribal or religious differences. Anyone who was white was considered European. The Company had to have special seminars for incoming managers and supervisors who had not worked with people of color before.

Ju Ju, more commonly known to Americans as *Voo Doo*, fascinated me from the beginning. Doc's steward and cook, his man 'Friday', who was named Sunday, began to suffer headaches, was becoming weak and eventually could not work. Doc was concerned and tried to get him to go to the hospital but Sunday finally told him that he had been cursed by an enemy and a witch-doctor had put a chicken feather on his doorstep that meant he would die. Doc refused to believe this, talked to some of the British, old West-Africa hands, and was told the man would die unless the curse could be lifted by another witch-doctor. They explained that his fear affected his liver. A Nigerian in the Company found another witch-doctor who was able to convince Sunday that he had overcome the 'curse' and Sunday believed him. Sunday eventually became the cook for Price, who replaced Doc, and Louise, who lived down the street.

Nigerians accepted that death was always near at hand and something you ignored. When it happened to somebody else, one felt good because it hadn't happened to you. Going to Saturday movies with the boys, we were amazed at how the Nigerians would laugh at scenes that would scare or sadden us. A neighbor with the Company, Marvin Redfern, whose house fronted Five-Cowry Creek, heard his wife scream one morning upon seeing a body washed up on their lawn. Marvin drove to the police station and several hours later they arrived, turned over the body, waved goodbye to his wife and left. They finally removed it several days later. Another day while in her yard, she saw a canoe pass by with several Nigerians, one stood up and relieved himself, and she shouted "Stop that!" which was a bad move. After that every canoe that passed repeated the incident.

By this time we probably had fifty expatriates and double that number in Nigerian employees. We had begun to renovate a new building on the mainland to move our office. Sarumi was grateful for the good business that had come his way and wanted to show his gratitude. He hosted an afternoon party with feasting for the entire Company. Upon our arrival, we were dressed in traditional Yoruba robes complete with hats in green and white, colors of the Nigerian flag, Libby and all expatriate women were dressed in their robes by the Nigerian women, including headdress. These robes were gifts from Sarumi that we still have. I brought my 'talking drum' and the Nigerians were thrilled with my drumming but my repertoire consisted of the familiar few notes used by NBC, the Nigerian Broadcasting Company, that we often heard on television. That began a hilarious conversation between the drummers in the band. We never knew what they said but they enjoyed themselves immensely.

Talking drums at Sarumi's party.

This was definitely a change from the traditional Saturday afternoon British curry luncheons that followed a morning of golf or other sport at the Ikoyi Club. We always enjoyed the informal 'formalities' of the British that had enabled them to remain 'British' in foreign countries for so long, while retaining health, dignity, individuality and wit. The Brits had a saying, 'Beware of the Bight of Benin, few come out where many went in'. Benin was the name of a river halfway between Lagos and the Niger River Delta on the coast, 'the slave coast', where the oil production had been found.

36. Libby: Paul's emergency and Teef Man

Before we arrived in Lagos, the whole family had to take pills daily for several weeks that were to protect us from malaria. We never failed taking the pills but both Eric and I still had what they called 'breakthroughs'. Eric got pneumonia at one point, not uncommon in tropical climates, and was sent to Port Harcourt to the Shell-BP Hospital and returned with high praise for the hospital, doctors and nurses there. A few months later, he had a breakthrough of malaria and we suspected that he had not truly recovered from the pneumonia. The breakthroughs were debilitating with fever and accompanied exhaustion, but unlike the real thing, only lasted for a short while. In Lagos, most European adults seemed to have something take them down at one time or another, but except for a few incidents, all the children seemed to sail through.

I was not the strongest after my long siege with brucellosis and I continued to have bouts of weakness but with no fever. Some years later, the doctors in Australia determined it was a 'syndrome' and cured me completely with a few months of tranquilizers. In Nigeria, after the brucellosis was cured, I had gained weight but was still a good candidate to have a breakthrough of malaria. Fortunately, we were now quite settled in the neighborhood, and happy with Bright, Matthew and the rest of the servants when a Malaria breakthrough struck me. After several days of fever, I awakened one afternoon and was laying exhausted in a bed of sweat with my eyes closed in the darkened bedroom. The boys were in school and Eric at work, but I heard the door open quietly and barefooted footsteps approach the bed. Then a hand lightly touched my forehead and then dear, gentle Matthew tiptoed out of the room. I'm sure he then reported to Bright that the Madame's fever had broken.

It's surprising that only two medical incidents occurred with the boys that are worth mentioning. John dislocated his big toe at Judo and his British instructor, pulled it out, snapped it into place and said, "There you go, Mate!" Then Paul had a stomach ache that we thought was just an excuse to get out of school for a few days but then Doctor Jim B determined that he had appendicitis instead. Jim B, Paul and I flew down to Port Harcourt to the Shell BP Hospital that afternoon. They operated on Paul in the evening and Jim B told me that all went well. I stayed in a small hotel within walking distance of the hospital and the next morning I anxiously walked over with some books that I had brought to read to Paul. The nurses, who proved to be just as marvelous as Eric had said, told me he was awake and that there was also a man in his room.

When I entered the room, Paul, who was ten, was awake and I think wondering what the heck had happened to him. He just thought he had a little pain in his side and flying off to a hospital sounded like fun. The man had his back to us and so Paul and I talked quietly, I read to him, and did all the things a mother would do for a son under the circumstances. I stayed through lunch and then returned early evening. The man still had his back to us and I gestured to Paul inquisitively about him. Paul looked at me and shrugged. The doctors wanted Paul to stay two more days and I arrived the next morning with a little chess set and more books. The man was talking to Paul but when I entered, he immediately turned his back to us again. Paul tried to whisper something but I decided to find nurses in the hall and ask about this man who was so unfriendly. They told me he was a Palestinian, had been shot in a political confrontation and was brought here to recuperate.

I returned the next morning and we were scheduled to fly out on the Company plane in the afternoon. When I entered the room the man was sitting on the edge of his bed and I greeted him, hoping he was feeling better. He stood up and in perfect English said he wanted me to know about him. He had been the oldest in his large family, his father had been killed when the Israelis took over their property, and to support the family, he had worked in the desert on pipelines twelve hours a day when he was 12 years old. Obviously, at some point he had been educated.

I knew almost nothing then of the Israeli-Palestinian conflict. I had noticed the Koran by his bedside, which Paul told me later that he read whenever they weren't talking. This man then said he had wanted to greet me because he did not know that American women loved their children! Such a shocking statement. However, he had seen that it was true with my concern over Paul and how I had treated him. It's easier to understand such misconceptions since TV was still a dream for most people in the world. We Americans lived in some far away land and Africans had no reason not to believe what their governments had told them.

Paul was soon back in school and I returned home after shopping one afternoon and entered the house by the back door. It was during the period after lunch when Bright and Matthew, and all servants, had their time off and I had scarcely entered when the front doorbell rang. I saw a young, well-dressed Nigerian inside our grilled porch; I pulled back the curtain, unlocked the glass front door and asked if I could help. He replied "Oh yes, Madame, I have been walking and walking for hours looking for a job and I am so very thirsty. If you could give me a glass of cool water, I would be ever so grateful." Never let it be said I wouldn't give water to a thirsty man! I said 'Of course', left for the kitchen and returned quickly to find him inside the house and moving towards my purse which was on a chair where I had placed it when I unlocked the front door. I stopped, told him to step outside and he said, "Yes, Madame". Then I asked him to step to the door of our porch, which he did. I sat the glass down on a metal table by the door on the porch and quickly shut and locked our door. I watched while he drank the water, replaced the glass, left and turned down the street.

I was undecided what to do for several minutes but when I saw him disappear into a yard further down the road, I decided to drive to Price and Louise's house to warn her, since he was headed in their direction. I saw nothing on the way, told Louise what had happened, and just then he appeared and quietly tried all of her locked doors while we listened. He moved on down the street and we decided to go for the police who were just a few blocks away. The police were very cooperative, decided that two of them would take their bikes and enter at either end of our long Ademola Street and work their way towards each other. We returned to Louise's house and before long heard a lot of noise in the street. There he was with several Nigerians and the two policeman. Curiosity got the best of us and we joined the group.

He was defending himself in a loud voice, but then the policeman searched him and found about 50 pounds and a letter. The policeman began reading the letter out loud and it was from his family. They said they were so happy that he was going to give up his *"teeving* ways" and come home! The crowd roared "Aha! *Teef! Teefman!*" Which of course, he denied. The policemen asked why he was walking in the neighborhood looking for a job with 50 pounds in his pocket at this quiet time of afternoon when people were resting. The people shouted louder, some were jumping around, and he was becoming fearful. Without us there, who knows what would have happened. The policemen then said they would take him to the station and when Louise and I offered to drive a policeman and *de teef* in my car to the station, they said, "Oh no, we very proud to walk him down the street with our bicycles so everyone can see *dis teefman!*" Watch them, we did.

Three weeks of repeating this great story to one and all, Louise showed up at my house late one afternoon after playing bridge with some British ladies who were telling stories about a successful thief who had been working on Victoria Island. He sounded so like our *'teef'* that Louise wanted to go immediately to the Police Station and find out what had happened to him. There was an inebriated Irishman there in charge who went through all the papers for the whole week that our thief had been brought in, we describing all details. The Irishman had no records of such an adventure and then said, "Fifty pounds? Do you really believe they would bring him in with fifty pounds?" He just smiled at our innocence, shook his head, then opened his eyes wide and said, "I'm returning home in three months!"

One morning Harriet and I went out looking for dress materials at several Lebanese shops that carried beautiful cloths imported from Europe and Asia. The Nigerian women all wore robes, consisting of a simple tunic, or blouse, with a length of cloth wrapped around their waist, then tucked in, and another length of cloth wrapped around their head. Sometimes there was no tunic and the material was just wrapped around their bosom. Often you saw them rewinding or re-tucking their cloths. They wore beautiful materials in all colors, but mostly green and blue, and many robes had embroidered tunics. Most of we expatriates had tried to have dresses made with their famous embroidery but try hard as we could in the fittings, they could really never seem to get it right. I suspect it was because we wore 'bras' and that just didn't compute with the men who tried to fit us.

This store was on the outskirts of the city and we were alone with the owner mulling over our choices when we heard people yelling outside. We saw a young boy running and a crowd chasing him, catching him, and then began beating him while yelling, "*Teef! Teef!*" I couldn't stand it for long and made for the front door to do I'm not sure what! The owner ran around me to the door and refused to let me out, telling me that they would begin to beat me if I intervened. Fortunately, we soon saw the boy struggle away and they let him go.

37. Eric: Robertkiri Oil Field and Cameroon

My position was now formalized as Senior Exploration Geologist and I had five exploration geologists to work out the geology for 30,000 square miles plus. This meant an increase in pay, albeit small, and to my dismay, I found out that production geologists made more money than exploration geologists! I supposed that it was because the production geologists were working in fields that were producing income.

Dale Overmyer, an experienced Mene Grande geologist, was appointed the supervisor of the production development. In my new position I was able to focus more on the regional geology outside of our permit areas. We had an expanding data base from our own wells and from information trades from other companies. This permitted us to develop ideas and to evaluate new exploration plays. The last exploration well that I worked on was the second location for the swamp barge and was called the Robertkiri 1. It was just west of Shell BP's Bonny Export Terminal south of Port Harcourt, and was surrounded by the permits that they had retained. When we ran the electric logs I was astounded by what appeared to be 1000 feet of oil sand. We ran pipe and it became our biggest oil field in the Eastern Delta. Eventually it became a production hub for Gulf and Shell BP, and had been based on seismic data obtained mainly from GR&DC.

The new offices for production, drilling and exploration had been moved into a large building on the mainland. The offices were small and strictly utilitarian. Charlie remained with the accounting and services departments in Tinabu Square but Newby came with us. The third drilling rig had arrived and it was a semi-submersible, enabling it to be floated and easily towed into deeper water locations. The export terminal was completed, all the pipelines had been laid from the rigs to the production platform and finally to the export terminal. The oil was piped to the production platform from all the completed wells to be combined and separated from the gas. The oil was then piped to the export terminal and ready to be pumped into a shipping tanker. Okan Field was now producing income for the Corporation.

Price, who had replaced Doc, was our resident geophysicist and was responsible for our local seismic activity in collaboration with GR&DC in Pittsburgh. The success of this work was measured by the discovery of nine new fields drilled with only ten wildcats. Price, who was a near neighbor, and I, traveled to the new offices on the mainland together early every morning. This was an adventure. The sights, sounds, and smells were always changing and except for the smells, were usually humorous. We were never bored during our 45 minute drive. One afternoon when returning from the office, traffic was unusually slow, practically a walk. We didn't have effective air-conditioning and as we inched along, Price looked at me quizzically. I checked my shoes and no, it's not me, it must be you. We were going so slow that he could check his shoes and then we realized that the smell was coming from outside. There was a tanker-truck in front of us and we noticed a liquid was dripping beneath. Then we knew we were traveling behind and would continue to have to follow this 'Honey-Wagon' until we crossed the bridge into Lagos Island. The Lagos sewer-system went raw into the harbor.

At this time the Company decided it must celebrate our success with a party for the whole staff of expatriates and Nigerians, families included. We hired what later became the internationally famous Fela Ransom Kuti Hi-Life Band. A typical Louisiana style shrimp-boil was organized by the Drilling Department and they would need shrimp to feed one-hundred people. Given the Lagos sewer system, we flew in blocks of frozen shrimp from Louisiana. All the guests had to shell their own shrimp and as we expatriates dug in, we noticed the Nigerians were hesitant or wouldn't eat the shrimp, period. First, they didn't know how the peel the shrimp. They finally told us how dangerous it was to eat, and then we had to assure them that it was fresh frozen from the USA. There was some splitting up of tribal groups but generally all went well.

We planned some sports activities for after work and had enough employees for several basketball and softball teams. The government had built a sports field that included basketball courts as well as a softball field and although very few Nigerians were familiar with basketball, these tall guys were quick to learn. However, soccer was and is the national sport. Knowledge of these games spread fast and at first we competed interdepartmentally but eventually were playing against other companies.

There was also a Nigerian Army basketball team and as well as Syrian and Lebanese teams. The games could be rough, but mostly through over-enthusiasm or ineptitude. One time a poisonous snake got onto a court and we all ran. Someone had a camera and the picture wound up in the newspaper. The way that articles were written in the paper was a constant source of amusement for all of us. Even in the most dire circumstances, English could be twisted to bring a chuckle. For example, "A severed human head was found in the path between (such and such) villages. Foul play is suspected."

The weather offshore the Niger Delta was mild except for occasional rain squalls. The tidal range was small, the waters were shallow, and there were no large ocean swells. We couldn't ask for better conditions. I continued to work on the wildcats, the logging programs, and evaluations. I felt it necessary to run all the electric logs, or electric profiles, in the wildcats. These exploration holes provided the only opportunity to determine the lithology of the sedimentary section that we were drilling and these rocks were not cemented for the most part. It was just like drilling in a sand-bank under water, which the drillers called 'snow-bank drilling'.

The loose sand and compacted mud would practically dissolve in the mud stream that we used to drill with. My feeling was that we needed to use all the electrical tools possible to distinguish the lithology from the drill cuttings. Once a month Newby would bellow my name from his office at the end of the hallway and I knew that the bill from Schlumberger had arrived. I had to explain again and again why I felt it necessary to use all of these electrical tools. It was the only way to obtain information on the thickness and extent of the oil-bearing sections to determine the possible quantities of oil we could produce. Newby knew that, but he wanted to hear me explain it over and over since part of his job was to reduce our costs to the minimum. He was known to repeat that, "oil was only worth two dollars a barrel IF you can sell it". Once Newby passed on a letter to me from the refining department in Pittsburgh that was looking for a heavier crude, a 25 degree bituman-based heavy oil, which brought a smile to my face. We had gained a reputation to be able to do anything.

I was beginning to realize that part of the success of Gulf Oil Corporation was by allocating great responsibility to local management in their exploration and drilling. In my overseas work, I had noticed that other companies took longer to come to a decision in their operations. I remembered how CalTex, with whom we were sharing a rig, had to have a meeting of both Boards before they could complete their well, which was costing them 25,000 dollars a day. Later I found out that Gulf had found more oil overseas in the 60's and 70's and at less cost than our major competitors. This was largely due to having experienced managers with authority.

Hollis, vice-president of exploration, requested that I be sent to Cameroon in the area adjacent to the easternmost boundary of Nigeria. I left with Mike K to search for a reported oil seep that had been written into the British Admiralty sailing instructions, the West African Pilot, in the nineteenth century. The letter said it was near the Man Of War Bay where British ships had found a fresh water supply. The ships had been sent there in the 1820s after Britain had passed a law against slavery to stop the slavers. This sounded far out to me, but Hollis had my confidence and respect. We had no detailed maps, just knew where the Bay was, period. Gene said if we really needed it, just let him know and he would send a helicopter.

Mike K and I flew to Victoria, the nearest town on the Bay with an airstrip. We found a decent hotel that used to be in the German part of Cameroon before World War I, had become British after the war, and the French territory was east of the mountain range. Now they were called the Cameroons. After WWII and independence, British Cameroon elected to become French because they did not want it to be part of Nigeria. Talk about confusion.

In the restaurant of the hotel, we attempted French but the waiter was uncomprehending but finally understood that we wanted *the book-for-look chop*. The next morning we rented a car and drove to the nearest place we could reach with a car on the Bay. There were no people, just an abandoned camp, that at one time was for promising students from West Africa. We found a leaky boat and rowed a couple of miles around the shore but found no sign of any stream of fresh water. We decided this just wouldn't work.

I called Gene and he said he'd have a helicopter there the next day that could carry three. We flew around the edge of the Bay for about five miles, seeing nothing except some workmen digging and carrying sand up the cliff. The pilot, Ian from Wales, found a landing space by the beach and I went over to talk with the workmen. I was finally able to communicate with one old man in 'Pigeon' English. I asked him if he knew of a place where there was 'kerosene', which we knew was their word for 'oil', he nodded and said, 'it be place where go whoosh, whoosh'. Kerosene was the first petroleum product widely distributed in Africa and had given them light and heat to cook. Their phrase for a superlative was 'sweet like kerosene'. Until kerosene was introduced, the highest mortality rate for children resulted from respiratory diseases from wood smoke. I asked him if he would show us the place in the helicopter and he accepted immediately. Mike K stayed behind and this old man's young comrades stood open-mouthed as we flew off. Within a few minutes the old man indicated the area and

Ian landed the helicopter. There was an oil slick and a hissing sound where gas was escaping from a fissure in the cliff nearby.

An oil seep at the foot of Mount Cameroon.

We gratefully gave the old man money, Ian took him back and returned with Mike K. We collected samples of the oil that looked similar to our Nigerian oil, and water as best we could, but had no container for the gas. Mike K suggested that we set fire to the gas to determine if it would burn like petroleum gas, but I demurred. If it was, we would never be able to put the flame out and besides, we were on a 'recreational trip' and had no permission for any business. We returned to Lagos, submitted our report and received personal thanks for our efforts from Hollis. We eventually had offices and leases in western Cameroon. Hollis had been studying references to oil and gas occurrences in East and West Africa, and as a result of his studies the Company had oil production in Angola-Cabinda and Nigeria. The production in Angola-Cabinda in 2008 is comparable to Nigeria's and deep-water exploration continues.

38. Libby: David's Village, Mark's Party, and Loompi

One morning I heard David knocking frantically at the back door, and Bright asked me to talk with him. David said that he had just heard that another village was being built next to his and he must go home to protect his children. When questioning him, I discovered, to my horror, that a new village was often consecrated by the spilling of a child's blood, who was then buried beneath where the village would be developed. He had just heard this news from a friend about his village. The thought of driving onto the mainland and into the jungle didn't appeal to me but I couldn't refuse. David was not only serious but was a nervous wreck, muttering in his language, and was in the car before me. After we crossed the bridge and were on the mainland he directed me onto some dirt roads. The trees became thicker and taller until everything was in dark jungle shade and it seemed forever until we arrived at his village.

Everyone turned out to see this little red car with the fair-haired driver. Children ran out of their huts, screaming and running after the car. David finally told me to stop and wait for him to return and during that period I became a real curiosity for all.

The children reached into the car to touch my hair and when some women tried to shoo them away, they only laughed and returned, and so I leaned my head out of the window for them to feel my hair. I remembered how surprised I was to feel such wiry hair when I gave a fond pat on the head to one of Doris's boys. Next they tried to touch my skin, so I put my hands and arms out the car window for them and couldn't keep from laughing while they felt and rubbed my skin, jabbered and giggled, their teeth so white with the sticks they used to clean them. These sticks used by all Nigerians were of a kind of wood that frayed with chewing and cleaned between their teeth.

African children were so happy and delightful to be around despite their poverty, compared with Bolivian children, who were serious and suspicious. I'm sure this was partly the result of the difference in the way the Spanish land owners had continued to control the land and enslave the people in Bolivia through the years, whereas in Nigeria, the British had not claimed the land, but just governed. Finally David returned, said he must stay and probably could not return to work for a few days, but that an old man on a bicycle would lead me back out to the main road. I waved goodbye to the children who screamed and yelled in their language and 'Goodbye' in English, and ran following the car through the village.

One afternoon I told Bright I must go upstairs to our air-conditioned bedroom to write letters and I didn't want to be disturbed. It wasn't long before there was a gentle knock on the door and Bright said, "Excuse me, Madame", but that Steven, Harriet and Jake's cook, wanted to talk to me. I knew Harriet had gone to the States and said I would talk to him later, but Bright said, "No, Madame, he must see you now", and at that, Steven stuck his head into the doorway. He began by apologizing for disturbing me and then, "I so sorry, Madame, my wife! Her waist pains!" I didn't know what he meant, Bright had retreated and Steven spoke poor English. I kept pushing for more information with no result when Bright stuck his head back in and said, "She is having a baby, Madame". I said I was happy for them and finally Bright said, "Right now! Madame!" and Steven queried "Hospital, please, Madame?" I grabbed my purse and followed Steven downstairs.

Outside, his wife, Josephine, was on all fours on the cement driveway. I opened the car door, and Steven started to push her into the back seat of our two-door Mini-Minor until I said, "No, YOU get in the back seat." There was road construction on Owolowo Road to the Hospital and big bumps that could throw my little car around. Josephine spoke no English, and I asked Steven if I should go fast or slow over the bumps and he said, "Fast Madame!" and about that time her water broke. The hospital had a very large elongated oval-shaped road to the entrance and when I roared up, Josephine struggled out and hobbled after Steven. I watched a minute, drew a deep breath and slowly drove out. However, before I completed the circle, here came Steven on the run. He had forgotten his hospital card and they wouldn't accept her without the card! Well, we raced back for his card and then raced to the hospital again. This time I waited to make sure all was well and in a few minutes, Steven ran out again, smiling, to say he had a son. On my return I cleaned the car myself, not daring to ask Bright or Matthew to do such a task.

When the Company decided they wanted to have a 'shrimp boil' for all the employees to celebrate the incredible oil finds, the bachelors, of whom we had many, asked me if I would make the sauce. I said I didn't have the spices required or any large containers but Newby gave me carte blanche to buy anything I needed. Somehow we found boxes of imported canned tomato sauce, they flew in some spices and I bought two green, enameled, ten gallon containers. When the time came, I began mixing in the kitchen and Bright just shook his head. He tried to convince me that Nigerians wouldn't eat shrimp, but after all, I had been ordered by the general manager. I insisted that some of the Louisiana drillers come try the sauce and they claimed it was excellent, but I knew later, after we moved to New Orleans, that it was not hot enough.

It seemed like every day was a celebration for Nigerians. They had radios that played music all morning and late afternoon. It wasn't unusual at all to look out our doors and see them dancing on any piece of cement they could find. Late afternoon was the time the traders generally came by so they could find the 'Master' at home. Once you bought something, you were forever on their list. Some Americans bought so much stuff that I'm sure they planned on going into a business at home. There were extremely hard, black woods as well as some soft woods that had been carved into figures, furniture and masks as well as antique brass pieces and beads galore, but there never seemed enough time for me to do a historical study of these pieces, which I regretted later.

One of the managers was leaving and spread the word that he wanted to sell his boat. Eric had never been keen about owning one, given all the possible problems, but when the boys overheard us talking, we couldn't resist their excitement and promises to help. On only our second time out on the boat the bloody thing began to leak heavily while we were in Lagos harbor. We bailed like crazy and somehow got safely back to shore. Although we were all were good swimmers and had life jackets, not one of us wanted to swim in Lagos harbor, full of raw sewage and the occasional shark. We never had it repaired and just made sure the person who bought it knew its condition.

All the boys had many friends and when Mark turned seven we knew we must have a birthday party. Large harbor boats were always available to take people to beautiful beaches on the mainland and it was a popular weekend trip with children. We decided on a beach party but would only take boys who were good swimmers and if parents agreed to this adventure. We had always tried to have just the number of guests as the years of the birthday celebrant, and this was no exception. However, both John and Paul wanted a couple of buddies too, and we thought this a good idea, making fourteen. We got to the dock early morning with baskets of sandwiches, potato salad, cokes and cake, had a fun ride over, found a lovely isolated spot and set up a buddy system so that each smaller boy had an older and bigger 'buddy'. The idea was to swim first, have lunch, and maybe a swim after lunch if they were willing to wait an hour.

Some boys were splashing around, others playing with sand and a few swimming out from shore when we saw a man at some distance running down the beach towards us. When he was in shouting distance we heard the words 'rip tide'. I cannot describe the terror that gripped us. Only four boys were in deep water but until we got them in, we had called on God many times. The boys had a marvelous time, consumed all the food, but agreed to return home after lunch. To this day Eric and I shudder in memory of that birthday party.

About this time a terrible thing happened. There was a bitch in heat across the highway, Loompi was crazy with love, slipped through Bright's feet and out the kitchen door, tried to cross Owolowo Road but never made it. Everyone was crushed. Bright felt terrible. It was late afternoon and news spread fast among the children. Eric was soon home and emptied out the tools from his wooden tool box, a scarce item in Lagos, he laid Loompi inside and hammered on a lid. We buried him under a banana tree and the boys built a cross for the grave which the little girls declared was necessary. When Bright saw the cross, he became very upset and said we must remove it immediately. He wouldn't even try to explain, saying that we couldn't understand but that it would bring bad *Ju Ju* to our house. All the children accepted his judgment, understanding that there was much about Africa that we could never know.

Bright somehow managed to feed us and got on his bicycle for home. After many tears, we tucked the boys in and then the electricity went out. Eric and I sat down with a candle and a bottle of scotch at the dining room table, but no amount of scotch made us feel any better. We went upstairs with our candles to check on the boys and found all three in one single bed, sound asleep with their arms around each other. Loompi had been with us since Bolivia, had sniffed the streets of Brooklyn and New York City, cities in Spain, France, and Nigeria as well as the mountains of Colorado. He'd bitten Mark's nose, had chewed holes in a couple of hand woven shawls and ruined one rug, but had been loved so very much.

39. Eric: Revolution and Transfer

Before leaving for our annual vacation in the fall of 1965, Gene informed me that I had been appointed to the Geological Committee of the Corporation and would have to attend a meeting in Harmarville, Pa. while stateside. This was a significant honor and I was pleased to meet other geologists in the Company that I had often heard about. There was much curiosity about how fast our operation had evolved and the magnitude of the reserves we had found. We had found more oil in Nigeria than Gulf had found offshore Louisiana in thirty years of exploration. Hollis, Gus and Mel were all at that meeting.

Shortly after our return to Lagos, Gene gave a party for visiting Gulf Oil Corporation officers, including the Executive Vice-President of Production and a member of Gulf's Board of Directors. In conversation with the Director, I found that nobody on the Board knew that we had a great deal of gas in our fields that had never been tested. He was shocked that he had not been informed of the amount of gas found, and was concerned that there was no immediate market. To my knowledge the export facilities in Nigeria have not yet been completed. Ironically, Gulf's lack of natural gas reserves, and a ruinous contract to deliver natural gas in the United States, at a fixed price, would contribute to the downfall of the Corporation.

At the time when I was beginning to feel that my work was completed in Nigeria, we had a Revolution. After school one Friday afternoon, we were returning Bruce Barrett, a good friend of Mark's, to his home. He lived on the mainland near the airport and it had been arranged that we would stop and see *Lawrence of Arabia* on the way, which had just been released, and we were to leave Mark at Bruce's house for the weekend. When Libby and I returned later that night, the streets seemed deserted, quiet, there were almost no candles burning at the stands, and truckloads of soldiers passed us driving towards the airport. This seemed unusual but we weren't concerned.

The next morning John and I drove with Price to the golf course where we had a golf date and John, now 13, had become quite a golfer, and was to play with some of his school chums. We got a late start at the course, waiting for a visitor to join us. He explained his tardiness as the result of a shooting that had occurred at the Ikoyi Hotel where he had just checked in the night before. Then a soldier had knocked on his door and told him to stay in his room. When he finally was able to leave, he saw blood all over the lobby and soldiers everywhere, but he didn't know what had happened. This was a hell of a welcome for the newly arrived manager of a Phillips operation.

We were on the fourth hole when the foursome behind us told us that a body had been found in the bushes near the military barracks that joined the golf course. We decided something dangerous was happening and returned to the Club House. Nobody was sure but there was suspicion that there had been a coup. We all had a beer, discussed the situation and decided we'd better leave. I found John who had not heard about anything, but I persuaded the boys to stop playing and we all went home. Bright only knew that there was 'government trouble' and that the Madame had gone out to buy some supplies.

We heard shooting that sounded like it was coming from the barracks across Owolowo Road. Libby returned with supplies that she felt might be necessary from unconfirmed stories that she had heard from neighbors on our street. Then we began to worry about Mark, although we knew he was in good hands. Without telephones we were at the mercy of rumors that told us the military was controlling the traffic on the road to the mainland and we were to remain in our houses. The next day we heard that a doctor, who was a friend of Bruce's family, was to bring Mark home when he could get through. It was at least another day before we found out that, indeed, a coup had occurred.

The Prime Minister, Sir Al Haji Abubakar, the father of Paul's friend, the man who had invited us to a reception for Julius Nyere, had been abducted and eventually was found assassinated, ritually dressed in his white robes and sitting under an old, large tree. He was highly regarded by all the people of Nigeria, but he was a Muslim and killed by the Christian Ibo officers from eastern Nigeria. They claimed that widespread corruption in the government, that was dominated by the Muslims, had motivated them. The coup was successful for only a short time before the Muslim officers took control and the Ibo officers fled to eastern Nigeria. Now there was no longer a democratic government. All of we expatriates felt uneasy and Libby and I couldn't help but think of the revolutions we had experienced in Bolivia.

On Monday we had a meeting of department heads and senior personnel, and were informed by the manager that he had been in touch with the American Ambassador. He was told that the situation was 'fluid', not resolved yet, but was assured that the leaders of the military government wanted foreign businesses to continue their normal operations. The Ambassador advised us to establish a central debarkation point where all of our people could be assembled to evacuate in the event security collapsed within the city. I heard later from Eddy Frankl, whom we had known in Bolivia, and had been at the Shell compound near Port Harcourt, that their dependents had already been sent out of the country before the coup. They had been concerned with the political disturbances in the East.

Within a few weeks, the Muslim officers, mostly Yoruba, had consolidated their control over the western part of Nigeria and our offshore operations would be the only source of income for this new government. Therefore we would be given protection from any offshore attacks. Commodore Wey, of the Nigerian Navy, came to our office. He apologized for the small size of his force but claimed he could protect our operations. He was a delightful, highly educated man, was embarrassed with what had happened and managed to make light of the situation. M.O. continued as assistant Minister of Mines and Power and there was no interruption in our business.

So much work had been completed in two and one-half years and I had begun to feel I was burning out. This development confirmed a decision that Libby and I had been tossing around, to return to the States. If we stayed, John would probably have had to enter High School in Switzerland that fall, we had been ten years in foreign countries, and the boys really didn't know their own country. I knew that a request for a transfer out of a foreign operation would not assure me of any preferential status in the domestic operations, but we really wanted to go home.

Gene didn't want me to leave but Libby and I were determined. When the word spread, a large party was planned for our departure and to welcome my replacement, Bob Johnston, who would come from California. It was a boisterous affair with dancing and good cheer all around. There was another party with the Nigerian staff to which the boys were invited, and they presented me with two very generous gifts; a large brass tray, about two and one-half feet in diameter, elaborately decorated and engraved around the edge with the words "To Eric K. Ericson from the staff of Nigerian Gulf Oil Company", and a large talking drum. These friendly people had worked hard learning a new business, were devoted to the Company and I greatly respected them. They had worked as hard as we did for the success of the Company.

The word came that we would be sent to New Orleans. An exciting city but not the climate of the west that we were hoping for. Perhaps they thought I could bring some of that Nigerian luck with me. We left for home shortly before the Biafran War officially broke out between the Ibos and the Muslim Government. The war effectively shut down the Shell BP operations, several unsuccessful attempts were made against our Escravos installations. Our house on Ademola Street where Bob and his wife were now living was damaged by debris from an exploding airplane that was flying towards Lagos to drop bombs.

As of 2008, the population of Nigeria has tripled, most city services are inoperable in Lagos, and expatriates and oil-workers generally need armed guards. Corruption is rampant and the people are very poor. There is tribal strife that never existed to this extent before. The oil brought prosperity but perhaps destroyed that democratic dream of one nation, that we had experienced in its very beginning.

The office says goodbye.

40. Libby: Biafran War and Going Home

The January coup had left us all unnerved, despite assurances from our Embassy and the Nigerian Government. When Mark finally got through the lines, over the bridge and home, with the help of a doctor who was allowed to travel, Mark told us about all the friendly soldiers. They had laughed, tossed his white hair, and even showed him how to hold their big rifles. Others had stories to tell about these 'friendly' soldiers. Our school stayed closed, we heard more shooting across Owolowo Road, and we were still advised to stay inside.

I had bought some beautiful plastic models for the boys in New York, of a human, a dog and a frog, with pieces that needed to be assembled, inside and out of the figures, with their bones, hearts, livers, and various parts, all appropriately colored. I decided that this was the time to use them. I can still see the boys studying the complicated instructions and completely engrossed over these models, taking time to help each other. It took days before they finished and then our curfew was lifted.

It wasn't long before Bright said that he must return with his family to Enugu, in eastern Nigeria. His aloofness had mellowed somewhat, he knew our eccentricities, had cared well for the boys and had trained Matthew to do most of his work, but I was still loathe to see him go. However, we understood his problem and never found out what happened to him and his family, or Matthew, for that matter, after we left in June. I had left Matthew with a lump in my throat. A year later, the eastern Ibo region seceded from Nigeria, proclaiming itself the Republic of Biafra, that resulted in Civil War. The war lasted about two and a half years, with over one million dead, mostly Biafrans, and various military governments, essentially Muslim, followed in Nigeria.

The challenges in Bolivia had basically been about survival, and Spain was mostly a cultural awakening for me. Those experiences helped guide me through the shifting values of the time as well as those busy and fast-changing years in Nigeria. Finally at thirty-seven, I begun to realize that there are limits to self-reliance, and I was ready for the security and comforts in the USA.

After all the emotional farewell fanfare we took a long road home. Five days in Cairo and Luxor, then Athens, Rome and Pompei, Copenhagen, an overnight train to Stockholm and Eric's delightful cousins, Spain and old friends, and finally to New York City, Kansas City and Boulder, Colorado, where family and friends were waiting to welcome us home. Such a long journey but a glorious adventure.

The Ericsons and Loompi.

Epilogues

Eric:

We arrived in New Orleans in July of 1966 and as for the climate, we felt like we were still in Lagos. I was soon in the Gulf offices that covered four floors in one of the central bank buildings in the city and found not one dark-skinned employee, quite a change. It was clear from the reception that I was a curiosity and they really didn't know what to do with me. I wound up with a small staff and was in charge of the exploration geology in the eastern flank of the delta where few fields had been found. This was ironic since if I had stayed with Continental Oil, I probably would have been in a similar position, however, the Gulf domestic organization was nothing like I had experienced with Continental in the Upper Gulf Coast.

The Mellon family of Pittsburgh financed the discovery of oil at Spindletop in Texas, near the Louisiana border in 1901, and this well opened up the Gulf Coast oil rush and gave the Company its name. Gulf's exploration and office in New Orleans covered the Delta, onshore and offshore, the contributor to the largest amount of oil and gas in their domestic organization. The competition for federal offshore tracts in the region was intense from all the oil companies, and communication between geologists and geophysicists was discouraged to minimize leaks of information outside of the Company. Few people had ever worked outside of that region and I felt that Gulf's geological education had been restricted.

The exploration manager invited me to attend the first federal offshore lease auction in several years; there were a large number of tracts, in and out of the production areas, and bids were high. Gulf did not bid on the more exploratory tracts and later in the office, the manager ridiculed Shell for their high bids. I suggested that Shell was concerned about oil supplies being interrupted after the Six-day War between Israel and the Arab countries in 1967, but they couldn't conceive that oil would ever be priced much over two to three dollars a barrel. Currently most of the oil production is found in countries in the Eastern hemisphere, which are unstable. The USA is the largest single consumer of oil, imports about sixty percent, and is faced with an unpredictable energy future. We must use less oil, and more efficiently, while we are developing new sources of energy.

I was asked by the general manager if I would help show a visiting official from the Nigerian government, who wanted to visit our Gulf operation, around the offices. I explained that it might be awkward since there were no African Americans in our office, even though half of the population of New Orleans was black. The visit was discouraged and he was a no-show. Not long after, Gulf hired its first professional African American whose skin was just slightly darker than mine. He was escorted and introduced through all the offices by the general manager. We later heard that President Johnson had brought together the chief-executives of the largest oil producing companies and told them that if they didn't begin hiring African Americans, they would not be allowed to participate in the federal lease sales.

After two years, in 1968, I was offered the job of exploration manager in Sydney, Australia, which would include exploration throughout the Southwest Pacific, including New Zealand, Tonga, Fiji, and Papua-New Guinea. All that exploration boggled my mind. I got several months of orientation at GR&DC, Pittsburgh, and I was also told that I would be responsible for the Gulf Rex, a geophysical research vessel that was on a world tour, when it would circle Australia and New Zealand. The family was just as enthusiastic about this new adventure as I was.

We had only been in Sydney for a month when the well in the Santa Barbara channel blew out and polluted some of California's most valuable real estate. Gulf's first exploration program in Australia had begun four years before, offshore, near the Great Barrier Reef. The spill in California raised the environmental awareness of the whole world and effectively closed our operation near the Great Barrier Reef where we had work obligations for over 50,000 square miles. As a result, I found myself participating in environmental forums and eventually as the principal witness for the industry in that vicinity for the Royal Commission inquiry that lasted for three years. They finally ruled against all commercial development in the 1,200 miles of the Northeast coast of Australia and made it the Great Barrier Reef National Marine Preserve.

Subsequently, our office participated in unsuccessful tests in Tonga and offshore Bougainville, north of the Solomon Islands. We also registered the Company for business in Papua New Guinea and ten years later oil was found in Gulf's acreage. In New Zealand we had to demonstrate that our vessel, the Gulf Rex, could operate without environmental risk. We agreed to travel with a witness from the environmental department who would make sure that we were not disturbing the annual tuna migration. The department was won over when our cook caught tuna off the stern while we were making geophysical surveys. About one-third of all oil produced in the world comes from offshore every continent except Antarctica. Reflecting on our travels and the different peoples with whom we worked, I am reminded that none of it would have been possible without peaceful coexistence between countries.

Libby:

Upon our arrival in New Orleans we were immediately met by old friends, from our university days in Boulder and those in the oil business. It was marvelous to be home again with clean food and water, dependable electricity, functioning cooling and heating systems, washers, dryers, dishwashers, telephones, and the infrastructure and services that government taxes buy; safe drivable roads, police, fire departments, libraries, and good public schools. And New Orleans, as Americans know, has more diversity to offer than the average city in our country. However, shortly after our arrival, we found that this was the first year for integration in the school system. Paul and Mark, still in grammar school, soon got into a fight while trying to protect a black girl that was being bullied by a couple of white boys during recess. They didn't know, and couldn't understand, 'prejudice'. We became active in the Unitarian Church that was teaching tolerance and bonded with our Colorado friends. When we finally thought we and the boys could handle this challenge of racism, Martin Luther King and Bobby Kennedy were killed, and all of that relief of being in the States began to fade.

Two years had passed when Eric came home, sat the boys down one evening and told them that he had been offered the job of exploration manager that would soon evolve into being manager for Gulf Oil in Australia, to be based in Sydney. Not even John, who would have been a Junior in High School that fall, wanted to stay, even though the family of his best friend offered to let him live with them. That was all the persuasion I needed. We were a close family. And after all, Sydney, was not even comparable to our other assignments.

Several months later, we flew through the States at Christmas time, visiting Jim and Doris in Washington, DC, family and friends in New York, Kansas City, Colorado and Los Angeles, flying into Sydney on New Year's Eve day, Eric claiming that this just might simplify taxes. Pants were just being accepted as 'proper' wear for women in the western world, and before leaving New Orleans, I had bought a royal blue, velvety corduroy pants suit with a padded jacket, swearing never again to travel long distances in a skirt and high heels. Upon arrival in Sydney, John and Paul were carrying my guitars and Mark his trumpet, and photographers thought us to be a rock band! What fun. We crashed for naps at the Wentworth Hotel, got dressed, and the concierge made reservations for a restaurant that the whole family could enjoy for New Year's Eve. We drove off in a taxi to a cove on the outskirts of Sydney Harbor and the perfect restaurant, everyone so friendly, with dancing, great food and family atmosphere. Then we asked for a taxi to go home. No luck. We had to begin walking about 1:00 AM, in this strange and enormous city.

Following directions, we walked up and down hills, passing drunks and broken bottles, and finally came to what looked like a hotel in King's Cross, the equivalent of Times Square in New York City. I told Eric I couldn't walk another step and was going to go in and sit or lay on that sofa I could see from the street. There were long sofas back to back in the lobby and the boys and I collapsed on one while Eric and John were outside with a mob trying to stop a taxi. Then I noticed an older man directly behind me with blood running down his neck. I stood up and asked if I could help and then he stood up, turned around and I saw a uniform and badges. He said "Ah'm right. Some bloke 'it me in me 'ead with a bot'le" and then asked what a 'Yank lady' was doing here alone in this lobby at this time. The boys stood up, we explained our situation, he announced that he was the Chief of Police and to follow him. He walked out into the middle of the street, raised his hand, stopped the first taxi, and to our embarrassment, made everyone get out and piled us in. Such an incredible introduction to an Australia that we learned to love.

We lived in Sydney for five years, John graduating from the Sydney Church of England School, Shore, returning to the States for university, and Paul from Sydney Grammar School. At that time, twenty-five percent of children in Australia went to private schools. Mark was in the fourth form or tenth grade at Grammar when we left Sydney. We have many fond memories of our years in Sydney, although I found the men to be quite sexist, a situation that has much improved with time. All of us communicate with friends in Australia, as well as with Americans and Canadians who lived there with us. But the Australian job and our foreign life finally ended when we were temporarily transferred to Midland, Texas. Our arrival back in the States coincided with the oil embargo of 1974 and we hoped that Americans would begin to understand their dependence on foreign oil and to make changes in their life-style. However, it has taken thirty-five years for that message to finally begin to sink in. We feel very grateful to have experienced our foreign life during those peaceful years abroad for Americans. We returned through Bangkok, Moscow, Paris and Spain. That life had been challenging, exciting and memorable, a period that shaped all of our lives with ever-changing views of our country and the world.

www.ingramcontent.com/pod-product-compliance
Lightning Source LLC
Chambersburg PA
CBHW020051170426
43199CB00009B/248